ELECTRON DIFFRACTION

ELECTRON DIFFRACTION

T. B. RYMER
M.A., Ph.D., F.Inst.P.
Senior Lecturer in Physics
University of Reading

METHUEN & CO LTD
11 NEW FETTER LANE LONDON EC4

First published 1970

Printed in Great Britain
by Butler & Tanner Ltd,
Frome and London
SBN 416 07660 2

Distributed in the U.S.A.
by Barnes & Noble, Inc.

Contents

8. ELECTRON DIFFRACTION BY GASES

9. ELECTRON INTERFERENCE

10. ELECTRON DIFFRACTION EFFECTS IN THE ELECTRON MICROSCOPE

APPENDICES

PLATES

Between pages 70 *and* 71

Preface

Undergraduate teaching usually relegates the subject of electron diffraction to an appendix to a course on X-ray analysis. This approach tends to under-estimate the importance of electron diffraction theory in the interpretation of electron micrographs, and of the practical technique for studying the structure of surfaces and molecules in the gas phase. Certain topics can be most easily understood in the context of electron diffraction. For example, an electron diffraction pattern is simply a plane section of the reciprocal lattice, whereas an X-ray pattern is a highly distorted representation. Thus the importance of the reciprocal lattice concept is most readily appreciated when approached via electron diffraction. Again, dynamical effects have only a small effect on X-ray patterns, but are beautifully illustrated in the diffraction effects which occur in the electron microscope.

The present book attempts to give undergraduate and postgraduate students a concise introduction to the subject of electron diffraction, with emphasis on general principles rather than detailed practical applications. For this reason, in discussing the dynamical theory, I have deliberately confined myself to the simple, two-beam approximation. I have also sought to bring out the connection between electron diffraction and other branches of physics by including chapters on electron interference and on diffraction effects in the electron microscope.

SI units have been used throughout. There is some doubt as to the appropriate unit of length for measuring a crystal lattice. The use of the Ångström unit (10^{-10} m) is being progressively discouraged, and spectroscopists are already beginning to express their measurements in μm (10^{-6} m) instead of the traditional Å. At the time of writing, there is little indication of crystallographers abandoning the Å. Yet it is not to be supposed that one branch of physics will indefinitely perpetuate a usage abandoned by the remainder. I believe that a textbook should be forward-looking in such matters, and I have therefore eschewed the Å. The possible alternative SI units are the nm (10^{-9} m) and the pm (10^{-12} m). The first of these has the disadvantage that the measurements of lattice parameters will usually have no significant figures before the decimal point. Measurements in the smaller unit, however, will usually have three significant figures before the decimal point, corresponding reasonably well with the accuracy of the majority of determinations. I have therefore expressed lattice parameters in pm. On the other hand, distances in the reciprocal lattice are most conveniently expressed in nm^{-1}, being usually in the range 1–10 in terms of this unit. Those unfamiliar with the newer units may find the following examples helpful.

$$1\ \text{Å} = 100\ \text{pm}$$
$$10\ \text{Å} = 1\ \text{nm}$$
$$1\ \text{Å}^{-1} = 10\ \text{nm}^{-1}$$

The unit cell of KCl is a face-centred cube of side 628 pm, and its reciprocal lattice is a body-centred cube of side 3·28 nm^{-1}.

The wavelength of a 50 keV electron is 5·5 pm.

I should like to express my thanks to Professor G. Honjo for providing the originals of Plates 5 and 6, and to Dr H. Wilman for providing the original of Plate 2.

Acknowledgments are also due to the following for permission to reproduce illustrations:

Verlag J. Springer (Plate 2)
Verlag Zeitschrift für Naturforschung (Plate 7)
Taylor and Francis Ltd (Plate 17)

University of Reading T. B. RYMER
February 1970

1

Historical Outline

1.1 *The wave nature of electrons*

The subject of electron diffraction originated in the experimental verification, by Davisson and Germer [16] and by Thomson and Reid [63], of de Broglie's [10] hypothesis that a moving electron should possess wave-like properties. This hypothesis, which is one of the foundations of quantum physics, is that a particle having momentum p can be represented by a wave of wavelength

$$\lambda = h/p, \tag{1.1}$$

where h is Planck's constant. Suppose an electron, initially at rest at a point where the electrostatic potential is zero, moves to a point where the potential is P. Then, equating the reduction in potential energy to the gain in kinetic energy,

$$Pe = p^2/2m, \tag{1.2}$$

where e and m are the charge and mass of the electron. Eliminating p from 1.1 and 1.2,

$$\lambda = h/\sqrt{(2meP)}. \tag{1.3}$$

Inserting the values of the constants h, m, and e into equation 1.3 gives

$$\lambda = \sqrt{(1500/P)} \tag{1.3a}$$

if λ is measured in pm and P in kV.

If the potential difference P used to accelerate the electrons is large, then the velocity of the electrons becomes comparable with the velocity of light, and it is necessary in equation 1.2 to use the relativistic expression

$$(m^2.c^4 + p^2c^2)^{\frac{1}{2}} - m.c^2$$

for the kinetic energy. The effect of this is that in equations 1.3 and 1.3a we must replace the actual potential difference P used to accelerate the electrons by the so-called 'relativistic' potential P^*, which is given by

$$P^* = P(1 + Pe/2mc^2). \qquad 1.4$$

If P, P^* are measured in kV, this may be written with sufficient accuracy as

$$P^* = P(1 + 10^{-3}P). \qquad 1.4a$$

The difference between P^* and P thus amounts to about 5% for $P = 50$ kV.

1.2 *Early electron diffraction apparatus: Low energy diffraction*

The experiments of Davisson and Germer [16] were carried out using low energy electrons (30–600 eV), which cannot penetrate more than a few atom layers into a solid. Any diffraction effects observed when a beam of such low energy electrons strikes the face of a crystal must therefore be due to a very few layers of atoms next to the surface. In Fig. 1.1, the circles represent rows of atoms

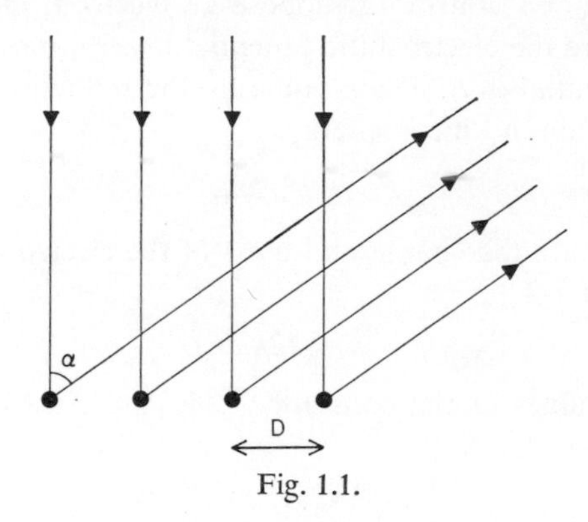

Fig. 1.1.

normal to the plane of the paper, and lying in the surface of a crystal. Let a beam of electrons be incident normally on the surface of a crystal, and be diffracted backwards through an angle α. The waves scattered by adjacent rows of atoms will interfere to give a diffraction maximum if the path difference is an integral number of wavelengths. From the figure, the condition for this is seen to be

$$D \sin \alpha = n\lambda, \qquad 1.5$$

where D is the distance between the rows of atoms and n is an integer.

Equation 1.3a shows that the electron beams used by Davisson and Germer had wavelengths ranging from about 150 down to 50 pm. Since the spacing of rows of atoms in a crystal is typically of the order of 150 pm, the angle of diffraction α will be at least 20°.

The energy of the diffracted electrons is too low to affect a photographic plate or Geiger counter, and the only simple method of detection is to collect them in a Faraday cylinder and to measure the current. The apparatus used by Davisson and Germer is shown in Fig. 1.2. A beam of electrons is formed by the electron gun A.

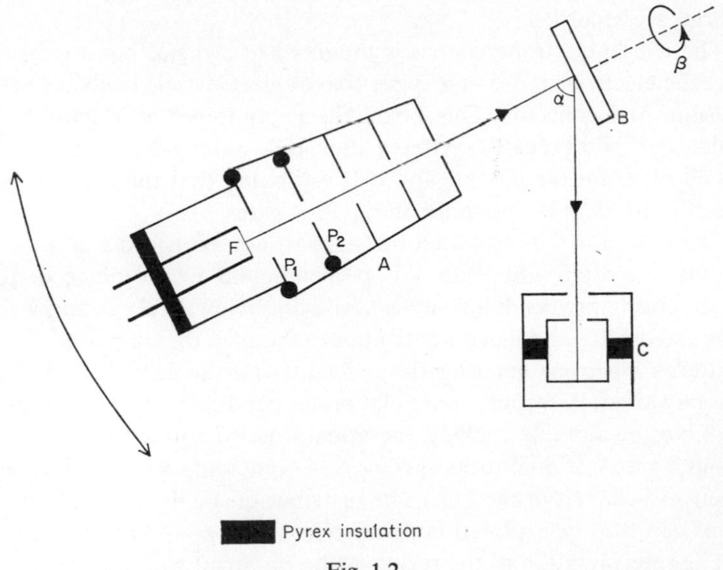

Fig. 1.2.

This consists of a tungsten filament F situated in a rectangular slot in a diaphragm P_1. The potential of this diaphragm is made slightly negative with respect to the filament. The resulting repulsive field between P_1 and the filament helps to concentrate the emitted electrons into a rather divergent beam. The electrons are then accelerated towards the diaphragm P_2, which is maintained at a fairly high positive potential, and some of them pass through a 1-mm diameter aperture to form a narrow, well-collimated beam. Three further apertures, maintained at a rather lower potential than P_2, provide a further collimation of the beam. These last apertures are each about 1 mm diameter and 8 mm apart. The beam of electrons from the gun strikes the (111) face* of a single crystal of nickel B situated about 10 mm away from the gun. The scattered electrons are collected in the double-walled Faraday cylinder C, which has an entrance aperture 1 mm diameter situated about 10 mm from the crystal B. The potential of the outer cylinder is the same as that of the crystal and the outer electrode of the gun. The inner cylinder is connected through a galvanometer to a point which is at a small positive potential with respect to the filament. This ensures that only electrons which have been scattered without appreciable loss of energy are recorded.

The whole electrode system is mounted in a metal box to ensure that the electrons move in a space free of electrostatic fields between the gun and collector. This box is itself contained in a glass bulb which can be evacuated. After a thorough baking out, the bulb is sealed off from the pumps, and it is estimated that the residual gas pressure inside it is then only about 10^{-8} torr.

The collector C is mounted on an arm, not shown in Fig. 1.2, so that it can rotate about an axis perpendicular to the plane of the figure and hang, pendulum-wise, vertically below B. By rotating the evacuated bulb and its contents about this axis, the direction of the scattered electrons entering the collector (the angle α of Fig. 1.1) can be varied. A second, more elaborate, pendulum system makes it possible, by suitably rocking the apparatus, to rotate the crystal B about an axis normal to its surface and coinciding with the incident electron beam from the gun. The distribution of the scattered electrons can thus be explored both in co-latitude (α) and azimuth (β).

The interpretation of the results to be obtained with an apparatus of this kind are dealt with more fully in Chapter 7, and only the salient points will be noted here. The energy of the electrons is determined by the potential difference P between the filament F and the main part of the apparatus; it is this potential difference which enters into equation 1.3a. We first consider the diffraction of very low energy electrons, when P is less than about 100 V. When the azimuth β of the crystal is adjusted so that the plane containing the incident and scattered beams is perpendicular to a

* For an explanation of the significance of the Miller indices (111) describing a crystal face, the reader is referred to any standard book on crystallography, such as that by Phillips [57].

densely packed row of atoms lying in the surface of the crystal, then it is found that for a fixed P the scattered electron current has a fairly sharp maximum at a particular value of the co-latitude α, showing that the scattered electrons form a well-defined beam. When P is varied, the angle α of the scattered beam is found to vary according to the equation

$$D \sin \alpha = \sqrt{(1500/P)} \qquad 1.6$$

formed by combining 1.3a and 1.5. A typical set of results obtained by Davisson and Germer is given in Table 1. The azimuth is such

TABLE 1

α (°)	85	80	75	70	65
P (kV)	0·0325	0·0340	0·0350	0·0365	0·0350
D (pm)	216	213	214	216	228

that the incident and diffracted beams lie in a (110) plane. With the exception of the last reading, the values of D determined from equation 1.6 using the experimental values of α and P agree well with the value 216 pm calculated from the known crystal structure of nickel. It is therefore reasonable to interpret the results as being due to diffraction of electrons by a *single layer* of atoms in the surface of the crystal.

If electrons of rather higher energy are used (P exceeding about 100 volts), then it is found that diffracted beams occur only for a narrow range of values of P. The explanation of this is that such electrons are able to penetrate into the crystal a few atom layers, and diffraction occurs from successive layers of atoms. Beams diffracted from different layers will interfere to give a diffraction maximum if the path difference between adjacent beams is an integral number of wavelengths. This requirement imposes a second relation between α and P in addition to equation 1.6. The two conditions can be satisfied simultaneously only for discrete values of α and P. The experimental data are in qualitative agreement with this interpretation, but a quantitative discussion requires a more refined theory than can be attempted here.

1.3 *Early electron diffraction apparatus: high energy electron diffraction*

Thomson's apparatus differed greatly from that of Davisson and Germer. He used values of the potential P in the range 10–60 kV. Electrons with this energy can penetrate solid films several hundred atom layers in thickness without appreciable loss of energy. They also blacken a photographic plate very easily, so that photographic recording of the diffraction pattern is very convenient. The wavelength of such electrons is less than 10 pm, and the angle between the incident and diffracted beams is therefore only a few degrees. These considerations led to the design of apparatus shown in Fig. 1.3.

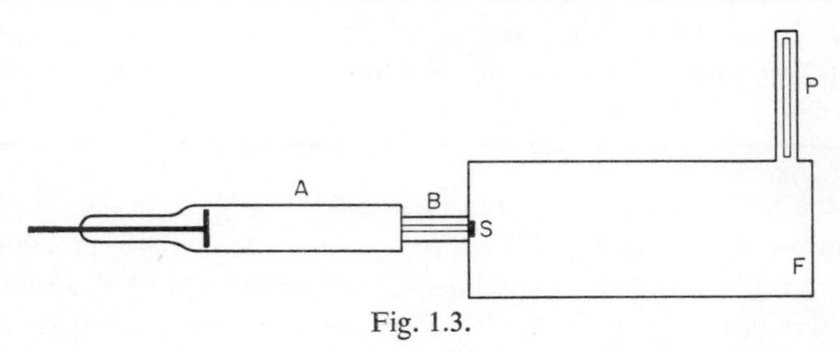

Fig. 1.3.

A low pressure, cold-cathode discharge tube A is the source of electrons. The necessary high potential is provided by an induction coil and measured by a spark-gap voltmeter. The electrons pass through a hole in the anode and are collimated by a narrow tube B, about ¼ mm diameter and 60 mm long. The collimator also serves to reduce the leakage of gas from the discharge tube (in which the pressure is about 10^{-3} torr) into the part of the apparatus to the right, where a pressure of 10^{-6} torr is maintained.

The collimator tube is surrounded by a thick iron tube. This serves to screen off the Earth's magnetic field, which would otherwise cause the electrons to move in a curved trajectory and therefore be unable to pass through the collimating tube.

On emerging from the collimator, the beam passes through the specimen S in the form of a thin film. At a distance of 325 mm beyond S, the electrons strike a fluorescent screen F, on which the diffraction pattern can be observed. Alternatively, a photographic plate P can be lowered in front of the screen when a permanent record is desired.

Electrons are diffracted by a crystal in the same manner as X-rays. Electrons falling at a glancing angle θ to planes of atoms with interplanar spacing d are reflected if Bragg's equation

$$2d \sin \theta = \lambda \qquad 1.7$$

is satisfied.* It is obvious from Fig. 1.4(a) that the angle between the incident and diffracted beam is 2θ. If, therefore, a beam of electrons is incident on a polycrystalline specimen S (Fig. 1.4(b)), the

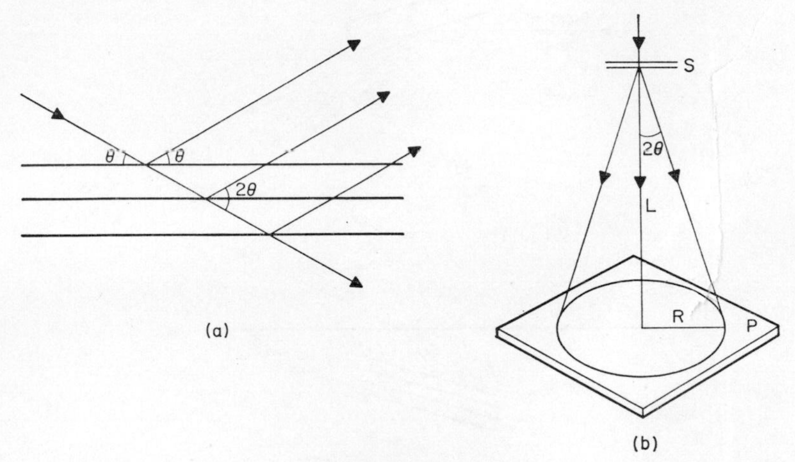

Figs. 1.4(a) and 1.4(b).

diffracted beams will form a cone of semi-angle 2θ, and therefore will fall on a photographic plate in a circle of radius R such that

$$R/L = \tan 2\theta, \qquad 1.8$$

where L is the distance between the specimen and the photographic plate.

We have already observed that the angle 2θ is only a few degrees when the accelerating potential P is of the order of tens of kilovolts. Hence in equations 1.7 and 1.8 we may write

$$\tan 2\theta \approx 2\theta \qquad \sin \theta \approx \theta$$

and then, on eliminating θ between these two equations,

$$Rd = \lambda L. \qquad 1.9$$

* See Appendix A.2.

The voltage P across the discharge tube was measured by a spark-gap voltmeter. The electron wavelength is then given by substituting the relativistically corrected voltage P^* (equation 1.4) in 1.3.

Measurement of the radius R of a diffraction ring, and the specimen-to-plate distance L, then enables the interplanar spacing d to be calculated. Thomson's early experiments were carried out with

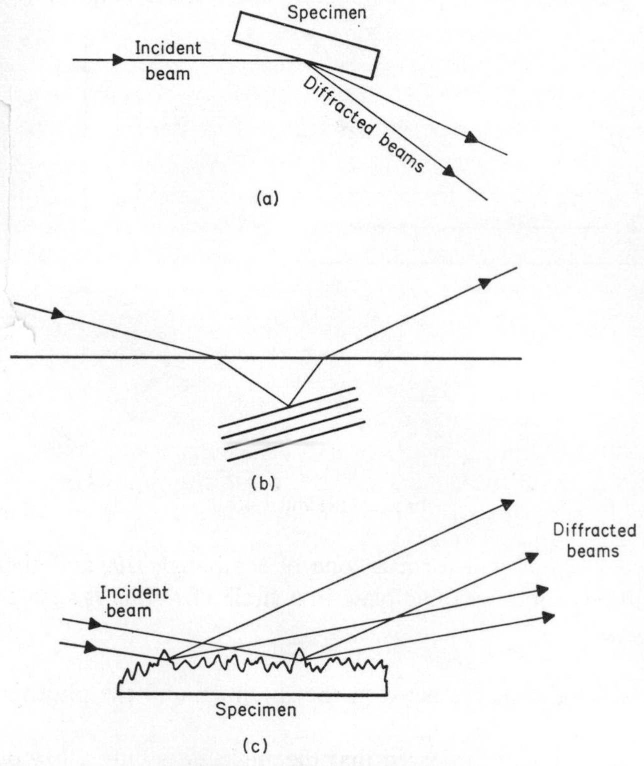

Figs. 1.5(a), 1.5(b) and 1.5(c).

specimens of gold, aluminium and celluloid, and the results agreed with the known crystal structures of these substances to within 2%.

It is obvious that an apparatus of this kind can be used for studying the crystal structure of specimens obtainable in the form of a film a few tens of nm thick. Such films, prepared by vacuum deposition or otherwise, are now of great technical importance, and their study is one of the main applications of electron diffraction. However, the scope of the electron diffraction technique was greatly extended by the discovery, by Nishikawa and Kikuchi [54], that diffraction patterns could be obtained using electrons reflected from the smooth surface of a specimen inclined at a few degrees to the incident electron beam (Fig. 1.5(a)). Further study shows that there are two types of reflection possible. If the surface of the specimen is *very smooth*, the electrons will penetrate the surface and be reflected by suitably oriented planes of atoms as indicated in Fig. 1.5(b). On entering or leaving the surface at a small grazing angle, the electrons are *refracted* through an appreciable angle. This is because the interior of a solid has a mean electrostatic potential about 10 V positive to the space outside, and therefore, according to equation 1.3a, the electron wavelength in the solid is less than outside. The origin of this so-called *inner potential*, and its effect on the diffraction pattern, are discussed in Chapter 5. Usually the effect of refraction at the surface is to distort the pattern badly and make it difficult to interpret. Under suitable conditions, however, it is possible to use the refraction effect to measure the inner potential.

If the surface of the specimen is suitably etched, so that the surface has numerous protuberances a few tens of nm in height, then the diffraction pattern is formed by electrons which have passed through the protuberances, as shown in Fig. 1.5(c). Since the electrons do not then enter or leave the solid at a small grazing angle, refraction effects are negligible. Diffraction patterns from bulk specimens are usually of this kind.

1.4 *Comparison of low and high energy electron diffraction*

The experiments of Davisson and Germer on the one hand, and of Thomson on the other, may be considered as signposts marking the two separate paths which the experimental development of the subject has followed. By far the greater part of the work on electron diffraction during the last forty years has used developments of the Thomson apparatus. The reason for this is that with this technique the diffraction pattern is usually formed by electrons which have passed through several hundred atom layers of material. An absorbed gas layer a few atoms in thickness will therefore not usually produce a significant contribution to the diffraction pattern, since its effect will be swamped by the much greater intensity of the diffraction from the underlying material. A relatively crude vacuum – about 10^{-5} torr – is therefore perfectly adequate for such electron diffraction work. It is merely necessary that the mean free path of the electron in the residual gas shall be large compared with the dimensions of the apparatus. The low energy electrons used by Davisson and Germer, on the other hand, are diffracted mainly by the surface layer of atoms. If the diffraction effects are to give information about the surface layer of the solid, and not about an adsorbed gas layer, the pressure of the residual gas in the system should not exceed about 10^{-8}–10^{-9} torr. It is only with the development, during the last few years, of demountable vacuum systems operating at such low pressures, that low energy electron diffraction has begun to compete in convenience with high energy electron diffraction. Now that convenient demountable, ultra-high vacuum systems are available, low energy electron diffraction is likely to become increasingly important. It is obviously the ideal technique for studying the structure of the surface layer of atoms of a solid, a subject of great interest in such fields as catalysis.

By measuring the position and intensity of the diffracted beams, it is obviously possible, just as in X-ray crystallography, to determine an unknown crystal structure. The use of electron diffraction, instead of the more usual X-ray diffraction technique, for this purpose has advantages in special circumstances. Electrons are diffracted much more strongly than X-rays. Thus, the Thomson apparatus used specimens about 10 nm thick and required only a few seconds' exposure, whereas in X-ray diffraction, specimens a few tenths of a mm in thickness are used and the exposure is usually measured in hours. As a result of the greater strength of the diffraction of electrons, it is possible to determine a crystal structure using less than 10^{-10} g of material. Some scores of new crystal structures have now been determined by electron diffraction.

The theory of electron diffraction is discussed in more detail in Chapters 3 and 6. A fairly accurate theory can be given of the intensities of the diffracted beams when high energy electrons are used, but it is very difficult to give a satisfactory account of the diffraction of low energy electrons. Consequently, determinations of crystal structures by electron diffraction are invariably made using high energy electrons.

1.5 *Diffraction by gas molecules*

We have noted in the previous section that strong electron diffraction patterns can be obtained, using a minute quantity of scattering material. It is therefore not surprising that a few years after the discovery of electron diffraction Wierl [66] applied this technique to the study of the molecular structure of gases. Wierl used an apparatus of the Thomson type, but instead of a solid specimen he used a

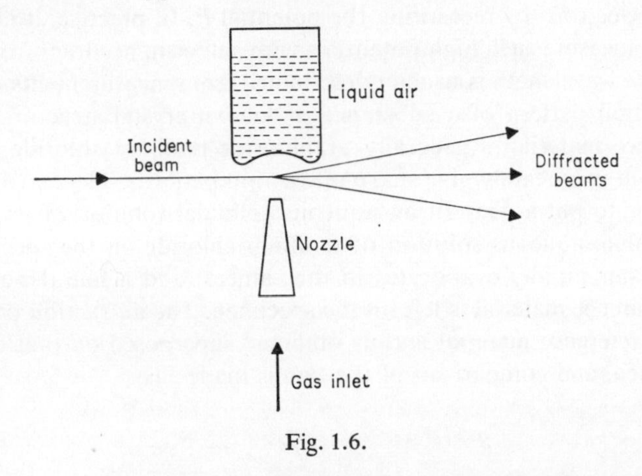

Fig. 1.6.

narrow jet of gas, which flowed across the electron beam and impinged on a metal surface cooled by liquid air, where it was condensed (see Fig. 1.6). The electron diffraction pattern obtained consisted of a number of diffuse haloes. From a pattern of this kind, it is possible to determine the various interatomic distances in the gas molecule. The structures of many hundreds of molecules have now been determined in this way.

1.6 *Wavelength measurement*

Unlike the characteristic X-ray emission lines used for crystallography, the wavelength of an electron beam is not fixed, but depends on the potential P used to accelerate the electrons. The early experiments of Thomson served to verify the correctness of equation 1.3. Later, more refined, experiments have aimed at using equations 1.3 and 1.4 to provide an accurate value of the quantity $\sqrt{(h/2em)}$. The wavelength of the electrons is found from careful measurement of the diffraction pattern produced by a substance whose crystal structure has been measured with high precision by X-rays. P is measured by means of a special high precision potentiometer. Experiments of this kind have been made by Rymer and Wright [60], Meyerhoff [46] and Witt [67], and are believed to have a precision of one part in 10^4 or better.

It is therefore possible in principle to determine the wavelength of the electrons by measuring the potential P. In practice, it is difficult to measure such high potentials with sufficient accuracy, and the electron wavelength is usually determined from measurements of the diffraction pattern of a substance of known crystal structure. The reference material is generally graphite or thallium chloride, both of which very readily give sharp diffraction patterns. The usual technique is to put a drop of an aqueous colloidal solution of graphite or a dilute aqueous solution of thallium chloride on the specimen. The water rapidly evaporates in the camera, and a fine deposit of the reference material is left on the specimen. The diffraction pattern of the reference material is thus obtained superposed on that of the specimen, and comparison of the two is made easy.

2

Modern Electron Diffraction Cameras

2.1 *Simple high energy camera: General description*

The diffraction cameras described in the last chapter are of historical interest only. In the present chapter, we shall describe some modern camera designs. The first great improvement in the Thomson camera was the introduction by Lebedeff [41] of a magnetic focusing lens. In the earliest cameras, this consisted of a short coil of wire with its axis roughly parallel to the direction of the electron beam; a steady current producing an excitation of about 2000 ampere-turns is passed through the coil. It is shown in textbooks of electron optics, e.g. Cosslett [13], that the effect of such a coil on the electron trajectories is very similar to that of a converging lens on rays of light. Figure 2.1 shows the electron trajectories in such a simple diffraction camera. The magnetic lens is represented schematically at L as if it were an optical lens. Electron beams from a source G are focused by L on to a photographic plate or fluorescent screen P. The specimen is placed at S. The undeviated beam is brought to a focus at C. A typical set of atomic planes gives rise to diffracted beams AD and BD which are deviated through angles CAD and CBD each equal to 2θ. It follows, from the geometry of the circle, that the points ABCD must lie on a circle, i.e. that the various diffracted beams are focused on a sphere of diameter approximately equal to the specimen-plate distance. Since, in practice, SP $\sim$ 500 mm and AB $<$ 1 mm, the depth of focus is such that the beams are sharply focused on the plane P perpendicular to the general direction of the electron beams. The replacement of the two fine collimating apertures of Thomson's original apparatus by a single aperture and a magnetic lens results in both a more sharply defined and a more intense diffraction pattern.

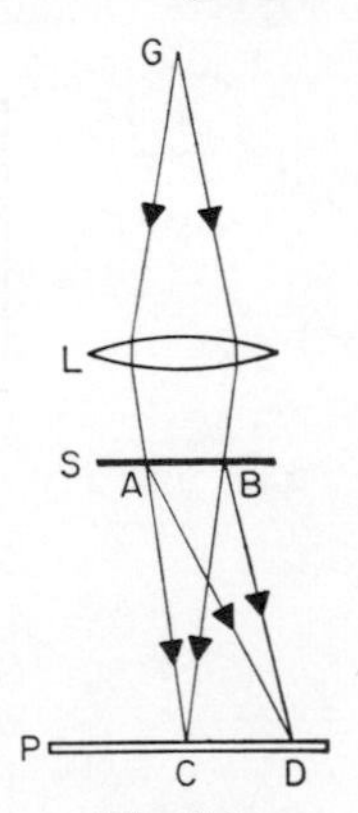

Fig. 2.1.

One of the earliest cameras using a magnetic lens was designed by Finch, Quarrell and Wilman [24]. Like Thomson's camera, this used a gas discharge tube as a source of electrons. The main body of the

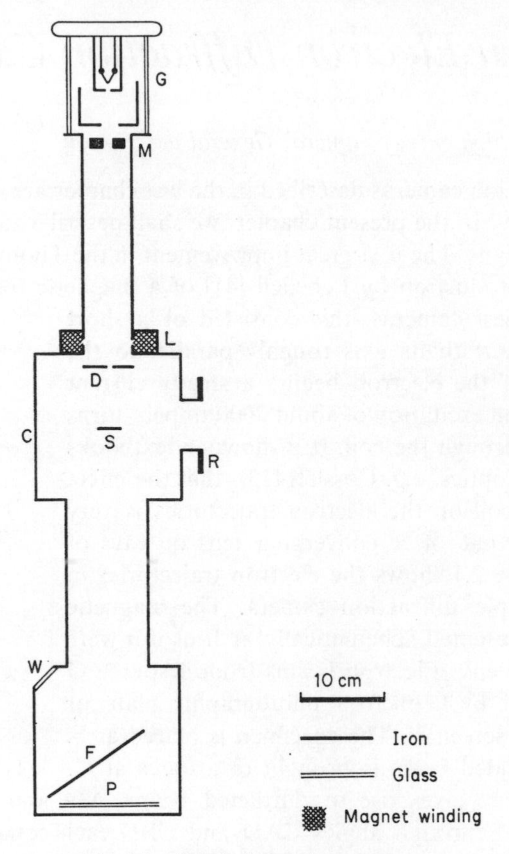

Fig. 2.2. Simple diffraction camera.

G Electron gun; M Beam aligning magnets; L Magnetic focusing lens; C Specimen chamber; D Diaphragm; S Specimen; R Vacuum port on specimen chamber; F Fluorescent screen; P Photographic plate; W Window. The fluorescent screen is shown partly raised.

camera was constructed of brass, and the magnetic lens consisted of a short coil, surrounding the camera body, excited with about 1500 ampere-turns. The vacuum system and high voltage electrical supplies were well engineered, and the camera was probably the most reliable of those in use at that time.

Later cameras have evolved from the Finch one, and the general arrangement of a typical one is shown in Fig. 2.2. The camera body is usually made of iron or steel, partly because this is a better material than brass for vacuum engineering, but mainly because a ferromagnetic camera body provides good screening of the electron beam from stray magnetic fields produced by other apparatus in the laboratory; such magnetic fields, by deflecting the electron beams in the camera, can distort or blur the diffraction pattern.

2.2 *Simple camera: Illuminating system*

At the top of the camera is the source of electrons G. Nowadays, this is an electron gun of the same type as that used in an electron microscope. This is more reliable and stable than the gas discharge tube used by Thomson and Finch; moreover, the absence of a gas discharge tube makes it possible to maintain a better vacuum in the main part of the camera.

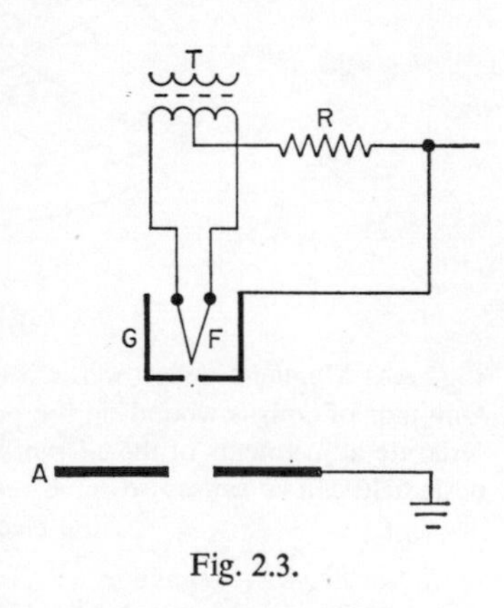

Fig. 2.3.

The essentials of the electron gun are shown in Fig. 2.3. The source of electrons is a V-shaped filament F made of pure tungsten wire about 100 μm diameter and between 10 and 20 mm in length; this is heated by a current from the transformer T having a secondary winding insulated to withstand the accelerating potential (up to 100 kV) which is applied between the filament and earth. The filament is sharply kinked at its tip as shown. This has the effect of slightly reducing the heat loss from this part of the filament, so that the temperature of the tip is slightly higher than that of the rest of the filament. Since the thermionic emission increases very rapidly with temperature, it follows that the bulk of the emission comes from a very small portion of the filament at its tip: in other words, we have a good approximation to a point source of electrons. The filament is surrounded by a closed cylindrical electrode G (variously referred to as the control electrode, grid, or Wehnelt electrode) having a hole about 1 mm diameter at its end. The filament tip is situated centrally in the hole and about 1 mm behind the outer surface of the electrode.

The control electrode is maintained at a steady potential of a few hundred volts negative with respect to the filament by the potential difference produced across the resistor R by the passage of the thermionic emission current from the filament. The electrons emitted through the aperture in G are attracted towards the anode A, which is at earth potential and about 10–20 mm distant from G. A high

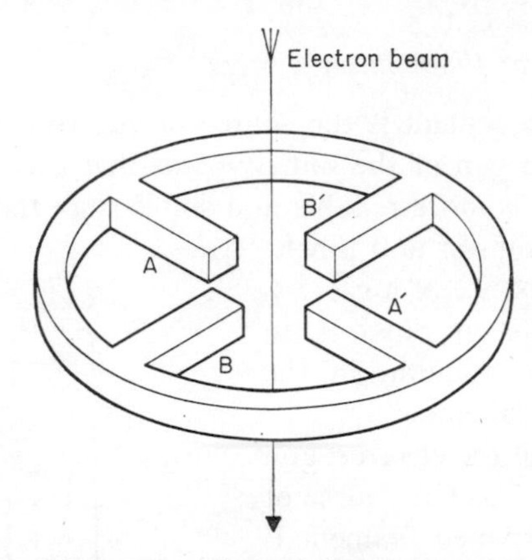

Fig. 2.4. Magnetic deflector for aligning electron beam along axis of camera. One pair of coils is wound on the pole pieces A, A′ and the other on B, B′. By separate adjustments of the currents in the two sets of coils, the resultant magnetic field can be caused to have any magnitude and any azimuth about the electron beam.

proportion of the total emission current passes through the hole in A, which is a few mm diameter, into the main part of the camera.

A gun of the type just described will easily produce a beam carrying a current of several microamperes with quite small divergence. Typically, the beam diverges to cover a circular disc of less than 10 mm diameter at 1 m. The exact dimensions of the gun are not particularly critical, the beam current and beam divergence being controlled by the filament temperature and the value of the resistor R. For details of the behaviour of such a gun, the reader is referred to a paper by Haine and Einstein [29].

In order that the beam emerging from the gun may be aligned along the axis of the camera, it is necessary that the filament tip should be accurately centred in the aperture of the control electrode G. A fine adjustment of the direction of the beam may be made by a small lateral displacement of the anode diaphragm A. Alternatively, a pair of electromagnets shown at M in Fig. 2.2, and in more detail in Fig. 2.4, can be mounted below A to produce a transverse magnetic field, adjustable in magnitude and direction, and extending over about 10 mm of the electron trajectory.

The electron lens L (Fig. 2.2) serves to image the source of electrons on to the photographic plate P or fluorescent screen F at the bottom of the camera. The general principle of its construction is shown in Fig. 2.5, which also shows how a vacuum-tight connection is made between the lens and the upper part of the camera; similar vacuum connections, not shown, are made between other parts of the camera.

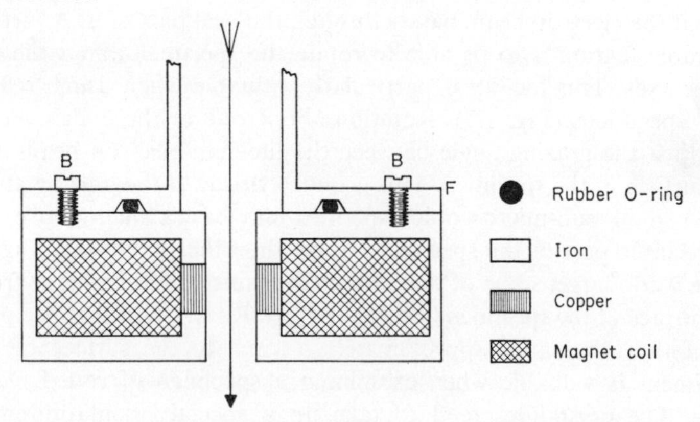

Fig. 2.5. A diametrical section of the magnetic lens. This is cylindrically symmetrical about the electron beam, which is indicated by a vertical arrow. The upper part of the camera terminates in a flange F, which is fastened to the top plate of the lens by bolts B. A rubber O-ring, compressed between the flat upper surface of the lens and a recess in F, makes a reliable vacuum-tight connection.

2.3 *Simple camera: Specimen chamber*

Immediately below the lens is the specimen chamber C (Fig. 2.2). This is often fairly large – 150 mm diameter and 150 mm long is not uncommon – so that auxiliary apparatus can be mounted in the vacuum, in order to measure other physical properties of the specimen while observing its diffraction pattern. The specimen chamber is provided with a number of ports, of which only one is shown in the figure at R. These ports serve for the attachment of various devices for manipulating the specimen, vacuum-tight closure being made with the aid of O-rings set in grooves in the surfaces of the ports in exactly the same way as the vacuum connection is made to the electron lens in Fig. 2.5.

The specimen S is mounted on a special holder (not shown in Fig. 2.2) which fits into the port R. The specimen holder is designed so that the position of the specimen can be adjusted from outside the vacuum system.

Ideally, it should be possible to move the specimen through a distance of a few millimetres in two perpendicular directions (which we may regard as the x and y axes) in a plane normal to the incident electron beam. This makes it possible to set a non-uniform specimen so that the electron beam passes through the best part of it. A further desirable feature is to be able to rotate the specimen about these x and y axes. This facility is particularly valuable when using 'reflection' specimens (Fig. 1.5). Rotation about one of these axes serves to adjust the grazing angle between the incident electron beam and the surface of the specimen. This is quite critical. If the grazing angle is too small, sub-microscopic asperities may cast a shadow over an appreciable part of the specimen. If, on the other hand, the grazing angle is too large, some of the diffracted beams will not emerge from the surface of the specimen (see Fig. 1.5(c)). Rotation of the specimen about the other axis, approximately normal to the surface of the specimen, is valuable when examining a specimen of rolled metal sheet. The crystallites tend to take up a special orientation with respect to the rolling direction, and therefore different diffraction patterns are observed according as the incident electron beam is directed along or perpendicular to the direction of rolling.

Numerous designs of specimen holder have been contrived which make possible some or all of the movements listed above. The rotary movements are usually made by control shafts passing through O-ring vacuum seals. Figure 2.6 shows a representative design. The reflection specimen Q is mounted on a plug P, which fits into a ring R. This is mounted in a holder H, and is fitted with a crown wheel C engaging a pinion W. The holder H is attached to a tube T which passes out of the vacuum system to a control knob A. A vacuum-tight seal is made with a rubber O-ring (shown in section in the figure as two solid circles) which is compressed in a V-shaped recess against the tube.

The electron beam is assumed to be directed on to the specimen along a line normal to the plane of the figure. Rotation of the knob A will therefore rotate the whole specimen assembly about an axis which lies approximately in the plane of the specimen and is perpendicular to the electron beam. Rotation of A thus serves to adjust the grazing angle of the electron beam on to the specimen.

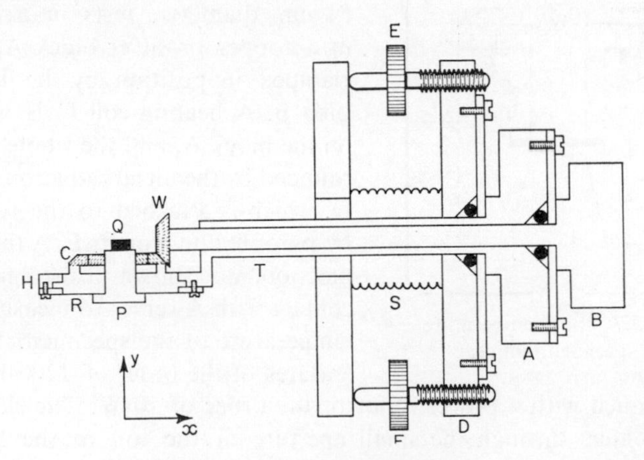

Fig. 2.6. Typical specimen holder.

The pinion W is driven by a shaft passing along the tube T to a control knob B. The shaft passes out of the vacuum system through an O-ring seal mounted in the knob A. Rotation of the knob B will obviously drive the gear wheels W and C and cause the specimen to rotate about an axis normal to its surface. This serves to adjust the azimuth of the electron beam in the specimen surface.

Translational movement of the specimen is usually provided by connecting the portion D of the specimen holder to the camera body by flexible metal sylphons. Two adjusting screws E and F are situated as shown. If these are rotated in the *same* direction, the sylphon will be extended or compressed, resulting in a movement of the specimen in the direction of the x axis. If, however, E and F are rotated by equal amounts in opposite directions, the plate D is rotated about an axis through its mid point perpendicular to the plane of the figure. This will cause the specimen to move in an arc of a circle, which is a sufficient approximation to a movement along the direction of the y axis.

If it is desired to study a specimen at high temperatures, it is possible to mount it in a miniature electric furnace attached to the specimen holder. The arrangement of a typical furnace for use with transmission specimens is shown in Fig. 2.7. This is based on a design published by Menter [45]. The specimen S, in the form of a disc about 5 mm diameter, rests in a cavity in a copper or silver block A. It is clamped in position by the flanged tube B. A heating coil C is wound on the block A, and the whole is surrounded by the metal radiation shield D, which is attached to the furnace by pyrophyllite mounts E. A thermojunction, not shown in the figure, in contact with A serves to measure the temperature of the specimen. Temperatures of the order of 1200°K can be attained with a power input of the order of 10 W. The electron beam enters through the small aperture at the top of the figure. The larger aperture at the bottom gives ample clearance for the cone of diffracted rays from the specimen.

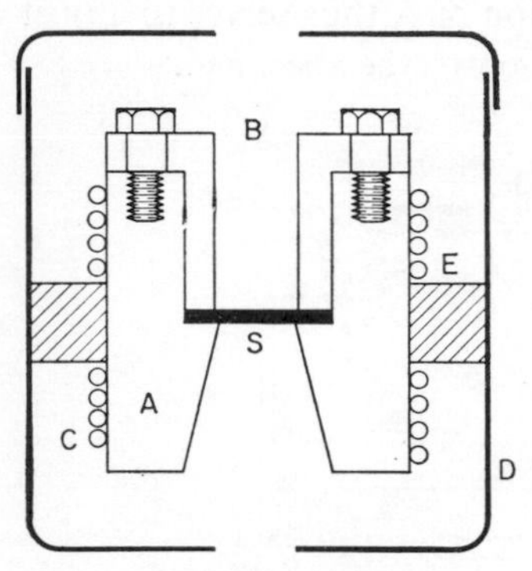

Fig. 2.7. High temperature specimen mount.

A similar device can be used for cooling the specimen below room temperature. Instead of using a heating coil, a number of 'Frigistors' are attached to the specimen mounting block. These are semiconductor junctions which produce a large Peltier cooling when a current is passed through them.

To attain the lowest temperatures, the specimen is mounted in a block attached by a massive copper bar to a special double-walled vacuum vessel containing a liquid gas, the vacuum in the space between the walls being continuous with, and maintained by, the vacuum in the camera. It is obviously difficult with this device to arrange for any means of moving the specimen, and its use is therefore restricted to specimens which are sufficiently uniform over a considerable area for the exact position of the specimen with respect to the electron beam to be immaterial.

This account of the contents of a typical specimen chamber would not be complete without reference to two other useful devices. The first of these is the specimen decharger. When examining by reflection (Fig. 1.5) a material which is an electrical insulator, it may happen that the specimen acquires a large negative surface charge from the incident electron beam. The resulting electrostatic fields around the specimen may so distort the trajectories of the diffracted electrons as to destroy the diffraction pattern completely. This trouble can usually be completely cured by introducing into the specimen chamber electrons with an energy of a few hundred electron volts. A suitable source of electrons for this purpose is shown in Fig. 2.8 (Picard, Smith and Reisner [58]). The filament F is surrounded by a control electrode G, and both are completely enclosed in the cylindrical anode A, which is in metallic contact with the camera body, being attached to the metal plate B; the last is mounted at one of the ports of the specimen chamber. G and F are connected through insulating vacuum seals S to electrical supplies which maintain F at a potential of about 500 V negative to the camera body, and G at the same, or slightly more negative potential. A diffuse electron beam carrying a current of the order of 1 mA is emitted through the holes in the anode A. This decharging device works by ionizing the residual gas in the specimen chamber. The resulting conducting plasma effectively conducts away any charge on the specimen surface.

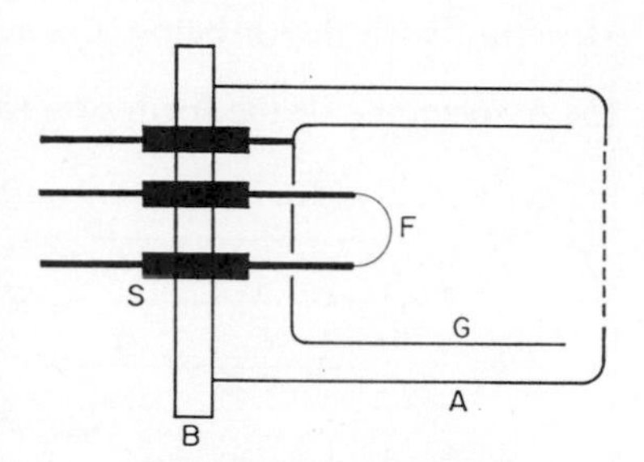

Fig. 2.8. Decharging gun.

Charging up of the specimen is much less serious with transmission than with reflection specimens, and it is seldom necessary to use a decharging device with the former. In particular, if the transmission specimen is mounted in a conducting tube of the general shape of the furnace block illustrated in Fig. 2.7, the incident electron beam will produce sufficient ionization in the confined space around the specimen to prevent any charge developing.

The remaining device which is sometimes used in the specimen chamber is an ion gun by means of which the specimen may be bombarded with argon ions having an energy of a few thousand electron volts. The following are two typical applications of this facility.

(*a*) Study of work hardening in a bulk specimen. A reflection diffraction pattern can be taken of the surface; this will be representative of the structure of a surface layer of material a few nm thick. Then, by bombarding the surface with ions, this thin layer can be removed and further diffraction patterns can be taken. Thus the variation of structure with depth below the surface can be studied.

(*b*) A specimen in the form of a foil can be thinned in the camera by

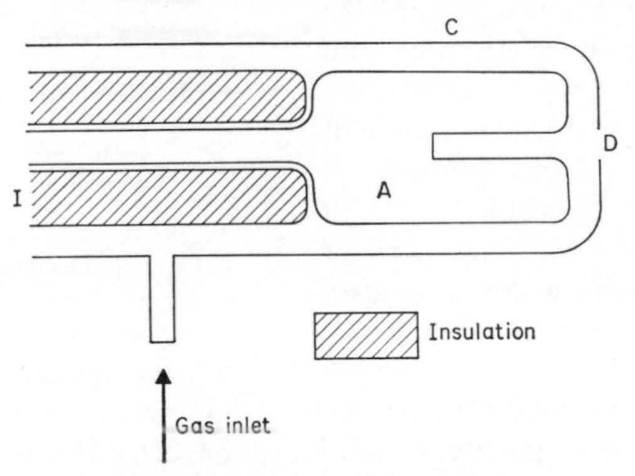

Fig. 2.9. Ion gun for etching specimen.

ion bombardment so that a transmission diffraction pattern can be obtained.

One of the simplest ion guns is that of Trüb Täuber [64], illustrated in Fig. 2.9. The anode A is mounted on an insulator I, and is surrounded by the cathode C. Opposite the hole D in the cathode from which the ions emerge, there is a deep cavity in the anode block. With this electrode construction, the bulk of the positive ions are formed from the gas in the cavity, and relatively few arise from sputtered electrode material. A suitable pressure of argon gas ($\sim 10^{-2}$ torr) is maintained in the space between the electrodes by admitting, at the point indicated, a slow stream of gas to replace that lost through the aperture D. With this gun, an ion current of about 1 mA at 3 kV can be obtained.

2.4 *Simple camera: Photographic recording device*

Below the specimen chamber is situated the fluorescent screen F (Fig. 2.2), on which the diffraction pattern can be viewed through the window W. A permanent record of the diffraction pattern can be made on the photographic plate P. A convenient, yet simple, arrangement is to mount the photographic plate in a light-tight box, which can be inserted into a vacuum port at the bottom of the camera. The fluorescent screen is attached to the hinged lid of the box, and can be lifted from outside the camera by rotating a shaft which passes through an O-ring seal. The pattern can thus be focused on the fluorescent screen, and then, by raising the latter, the photographic plate can be exposed.

More elaborate plate-holders have been devised which enable plates to be inserted into the camera and removed again through air locks, thus enabling a sequence of patterns to be taken of a specimen without the necessity of breaking the vacuum in the camera. An example of many such devices is the one described by Zworykin *et al.* [68]. Another way of obtaining a sequence of patterns is to use a roll film instead of a photographic plate. This alternative is less favoured because a roll of film takes a long time to de-gas, and therefore it is difficult to maintain a really high vacuum in the camera.

Whatever type of plate-holder is used, it is important to mount a small, earthed metal disc immediately above the plate, so as to intercept the main, undiffracted beam of electrons. Failure to do this results in the centre of the plate acquiring a large electrostatic charge. The trajectories of the diffracted electrons are then deformed by the resulting electrostatic field, with the result that the innermost diffraction rings and spots are displaced outwards in the pattern. The resulting error in the radii of the innermost diffraction rings can amount to about 0·2%.

2.5 *Simple camera: Kinematic recording*

The device known as *kinematic recording* is occasionally used to exhibit the change in the diffraction pattern of a specimen when some parameter (e.g. the temperature) is gradually changed. The principle of the method is shown in Fig. 2.10(a). The diffraction pattern of the specimen S is formed on a diaphragm D in which is a slit allowing only a narrow diametral strip of the pattern to pass through to the photographic plate P. The plate is moved perpendicular to the slit, as indicated by the arrow, in synchronism with the change which is being made in the parameter of the specimen.

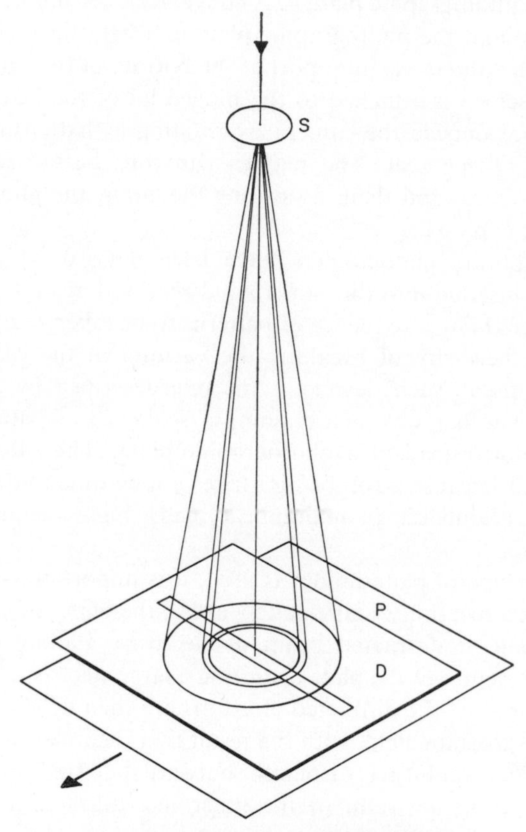

Fig. 2.10. Kinematic recording camera.
(a) Principle of camera.

Figure 2.10(b) illustrates the general appearance of the pattern produced. In the example depicted, it is supposed that the temperature of the specimen is gradually raised as the plate is moved, and that the material of the specimen undergoes a phase change from a hexagonal close-packed structure (corresponding to the pattern on the left) to a body-centred cubic structure (corresponding to the pattern on the right) at a temperature of 504°K.* It is obvious that the transition temperature can be inferred from the position on the photographic plate where there is an abrupt change in the pattern.

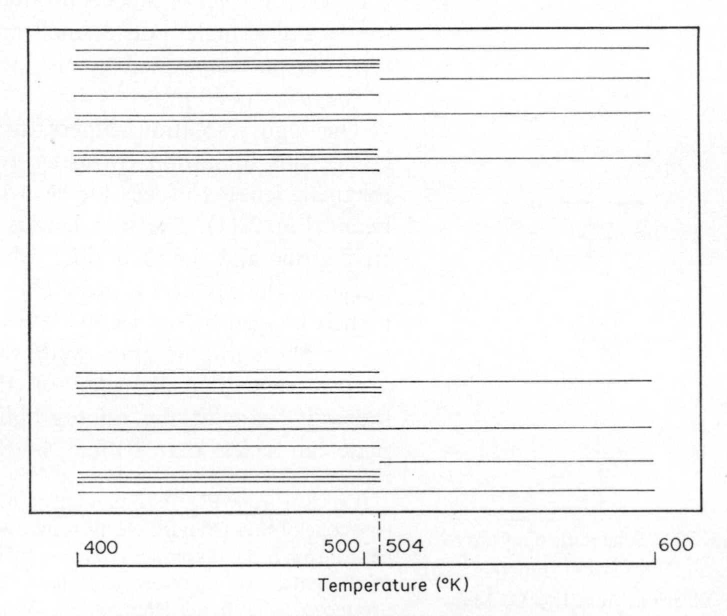

Fig. 2.10. Kinematic recording camera.
(b) Diagram of typical diffraction pattern.

2.6 *High resolution diffraction system*

One of the limitations of the simple system described above is the size of the focused spot at the position of the photographic plate. Since the magnetic lens behaves exactly like its optical counterpart, the size of this spot is equal to the size of the electron source multiplied by the ratio of the distances of the lens to the photographic plate and electron source. Unless the distance between the electron gun and the lens is made unreasonably large, it is impossible to reduce the focused spot size below about 50 μm when the specimen

* The patterns illustrated have been computed, but they refer to a real case – thallium.

to plate distance is about 500 mm. This makes it impossible to study such effects as refraction of the electrons at the facets of the sub-microscopic crystallites of which some specimens are composed (§§ 5.1; 6.1); such refraction effects produce, in place of a single spot, a small pattern of spots separated by distances of the order of 50 μm.

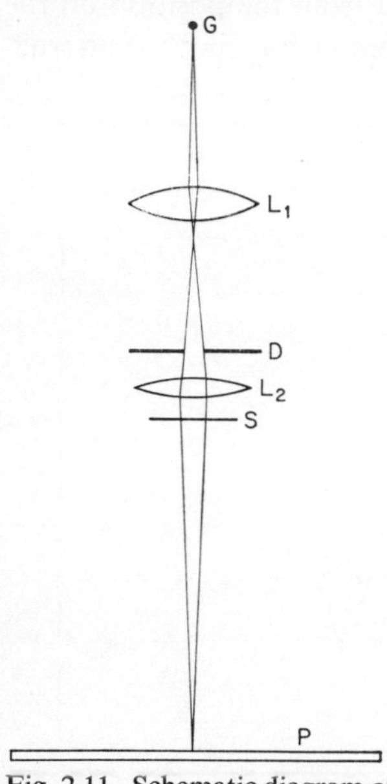

Fig. 2.11. Schematic diagram of high resolution camera.

G Electron gun; L_1, L_2 Magnetic lenses; D Diaphragm; S Specimen; P Photographic plate.

Typically, $GL_1 = L_1L_2 =$ 250 mm; $L_2P =$ 500 mm.

The high resolution camera overcomes this limitation by using *two* magnetic lenses to focus the electron beam (Fig. 2.11). The first lens is a strong one and forms a diminished image of the electron source. This in turn is imaged by the second lens on to the photographic plate. With this arrangement, the diameter of the focused beam at the photographic plate can be less than 1 μm.

It is clear from Fig. 2.11 that only a narrow cone of rays from the electron gun will pass through the diaphragm. The intensity of the pattern is therefore less than that produced by a single lens camera. While intensity is not usually a serious limitation in electron diffraction work, it is nevertheless convenient to use a single lens camera unless the high resolution facility of the double lens arrangement is essential.

2.7 *Convergent beam system*

If the strength of the magnetic lens is increased, so that the electron beam is focused near to the specimen, then, as will be seen from Fig. 2.12, the pattern formed at the photographic plate by the undiffracted rays is a 'pinhole camera' image of the specimen. It sometimes assists the interpretation of a diffraction pattern to be able to correlate it with a well-defined crystal shape in the convergent beam image.

It is clear from Fig. 2.12 that the direction of the electron beam is slightly different at different parts of the specimen. If, therefore, the specimen is in the form of a monocrystalline lamina of uniform thickness, the pattern formed by the undiffracted beams at the photographic plate will indicate the variation of the intensity of the

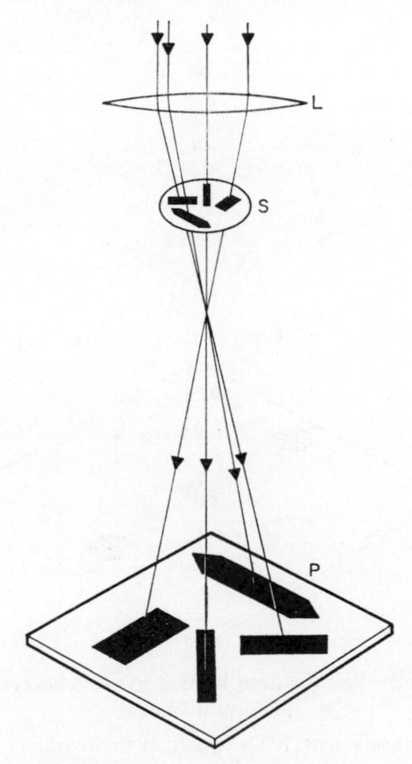

Fig. 2.12. 'Pinhole camera' image of specimen formed when magnetic lens is over-focused.

L Magnetic lens; S Specimen; P Photographic plate.

undiffracted beam with the direction of the incident beam in the crystal. Figure 2.13 shows that the diffracted rays will give rise to similar patterns on the photographic plates; any one such pattern will be a representation of the variation, with direction of the incident beam, of the intensity of the diffracted beam. The theory of such convergent beam patterns is discussed more fully in §§ 3.5 and 6.2.

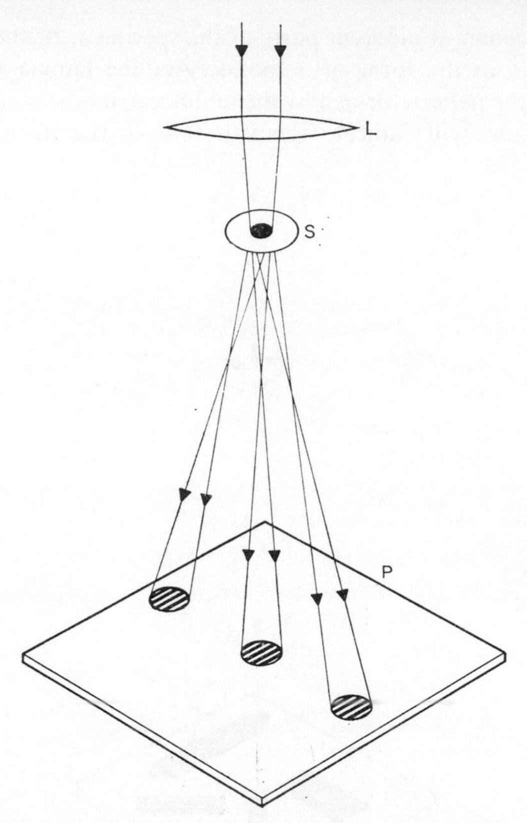

Fig. 2.13. Convergent beam pattern formed by a monocrystalline specimen of uniform thickness.

L Magnetic lens; S Specimen; P Photographic plate.

2.8 *Use of the electron microscope as an electron diffraction camera*

Nowadays, most electron diffraction work is probably done using an electron microscope. An electron microscope can produce excellent diffraction patterns, and the interpretation of these is often helped by the ability to correlate them with electron micrographs of the specimens. Moreover, it is possible to magnify the diffraction pattern by the electron lenses of the microscope so that fine details become easily visible. With a conventional electron microscope, it is easily possible to produce a diffraction pattern of a size corresponding to a simple camera some tens of metres in length. Since the cost of the simple type of camera described in § 2.1 is a considerable proportion of that of an electron microscope, there is little incentive to build a simple camera unless it is necessary to have special facilities, such as a very large specimen chamber, which are not included in the usual electron microscope.

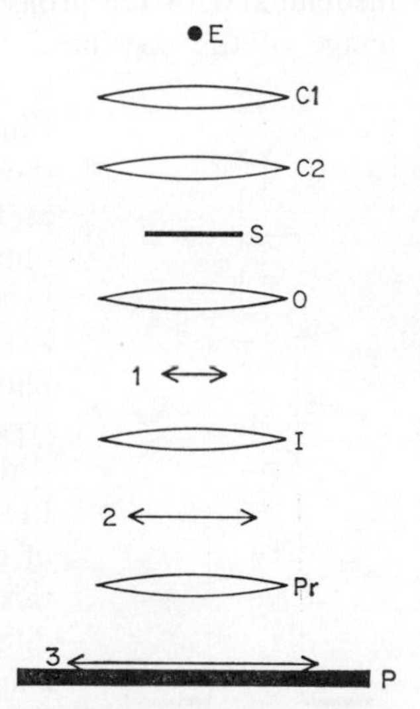

Fig. 2.14. Schematic diagram of electron microscope. The magnetic lenses, constructed as in Fig. 2.5, are here represented as if they were optical lenses.

The modern electron microscope has five magnetic lenses arranged as in Fig. 2.14. To produce the usual electron micrograph the condenser lenses C_1 and C_2 are adjusted to produce a narrow, intense, well-collimated beam at the specimen S. The objective lens O forms a real image of the specimen at 1. The intermediate lens I forms a real image of 1 at 2, and this latter image is in turn imaged by the projector lens Pr at 3 on to the photographic plate P. The reason for using *two* lenses I and Pr to image 1 on P is so that the overall magnification can be varied over a wide range. It is clear that if the strength of the lens I is reduced, the image 2 will move towards the projector lens Pr, and will become larger. If the strength of the projector lens is increased so as once more to image 2 on to P, it is clear that the magnification produced by Pr also increases. Thus, by reducing the strength of I and increasing that of Pr, the overall magnification produced by these two lenses is increased.

The usual way of using an electron microscope to produce a diffraction pattern is known as *selected area diffraction* (Le Poole [42]). The method can be understood by reference to Fig. 2.15. A fine, parallel beam of electrons (produced by the electron gun and two condenser lenses, which are not shown in the figure) is incident on the specimen S. The figure shows, emerging from the specimen, the undiffracted beam and two diffracted beams. These are each brought to a focus in the back focal plane, D, of the objective O.

In other words, a diffraction pattern is formed in the plane D. The figure shows how the electron rays travel on and form an image of the specimen in the plane I. The boundary of this image is formed by a diaphragm in the plane I. It is clear that by adjusting the intermediate and/or the projector lens, it is possible to focus either the image of the specimen, located at I, or the diffraction pattern, located at D. Further, since the only electrons which can reach the photographic plate are those which have passed through the aperture at I, it follows that the diffraction pattern obtained by this technique is that produced by the portion of the specimen whose image is seen when the plane I is focused on the photographic plate. Since the image at I has, typically, a magnification ×25, it is possible without using a particularly fine aperture at I to obtain a diffraction pattern from an area of the specimen only a few μm in extent. In this way, it has been possible to obtain a diffraction pattern of a single crystal from a specimen of a clay-type material.

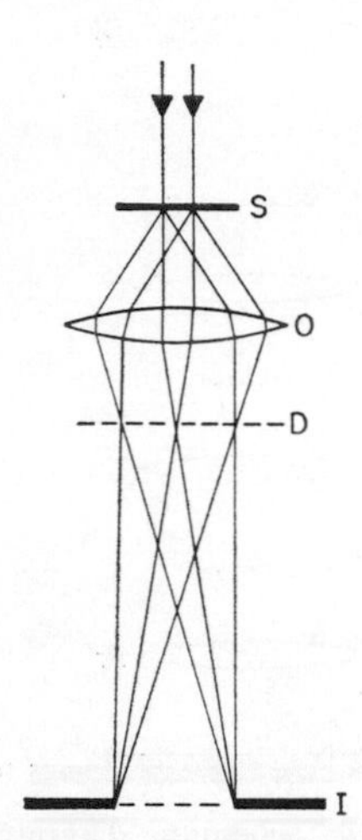

Fig. 2.15. Ray diagram of electron microscope used for selected area diffraction.

The limitations of the selected area diffraction method have been discussed by Riecke [59]. The principal sources of error are spherical aberration of the objective lens, and the difficulty of ensuring that the image of the specimen is situated precisely in the plane of the diaphragm at I. These effects combine to produce a discrepancy of rather less than 1 μm between the boundary of the portion of the specimen giving rise to diffraction pattern, and the boundary of the portion imaged.

A second way of obtaining a diffraction pattern with an electron microscope has been described by Bassett and Keller [5]. The two condenser lenses are used, in the manner of the high resolution camera described in § 2.6, to produce a diffraction pattern just below the objective lens, which is switched off. This diffraction pattern is enlarged and projected on to the photographic plate by the intermediate and projector lenses. If the pattern is slightly defocused, then one obtains a 'pinhole camera' picture of the specimen in the manner described in § 2.7 and illustrated in Fig. 2.12.

2.9 *Scanning electron diffractometer*

The instruments hitherto described in this chapter all use a fluorescent screen for visual observation of the diffraction pattern, and a photographic plate when a permanent record is required. In many ways, a photographic plate is an ideal recording medium for an electron diffraction pattern. It is a simple, cheap and highly sensitive detector of electrons (even a single electron can produce a perceptible effect); moreover, it records a two-dimensional distribution of intensity with high spatial resolution. It has, however, some disadvantages. In the first place, if it is desired to make quantitative estimates of the intensities of the diffracted beams, it is necessary to use a microphotometer to measure the blackening of the plate at the diffraction rings or spots, and thence to calculate the intensities with the aid of an intensity/blackening calibration curve. This procedure is time-consuming, and greatly detracts from the convenience of the photographic plate as a recording medium; moreover, the accuracy of intensity measurements made in this way is only a few per cent.

A further disadvantage of the photographic plate is that it has a logarithmic response to electrons, so that the rate of increase of blackening with electron beam intensity becomes small at high intensities. This means that weak diffraction spots or rings cannot easily be observed in the presence of strong diffuse scattering.*

The third reason why a photographic plate is unsuitable for accurate intensity measurements is that theories of diffraction refer to electrons which have been *elastically* scattered within the specimen. However, a high proportion of the intensity in an ordinary diffraction pattern is due to electrons which have undergone *inelastic* scattering, and these must be removed before meaningful intensity measurements can be made.

The scanning diffractometer is designed for accurate intensity measurements. The general principle is to use a simple diffraction camera, and to replace the photographic plate by a small aperture

* To some extent, this defect may be offset by a fortunate peculiarity of the human eye (§ 8.4) which makes it possible to measure quite accurately with a travelling microscope the *position* of such weak spots or rings even though it is not practicable to measure their *intensities* accurately with a microphotometer.

behind which is placed a device for measuring the current carried by any electron beam passing through the aperture. The diffraction pattern is scanned across the aperture by the use of magnetic deflector coils placed near the specimen.

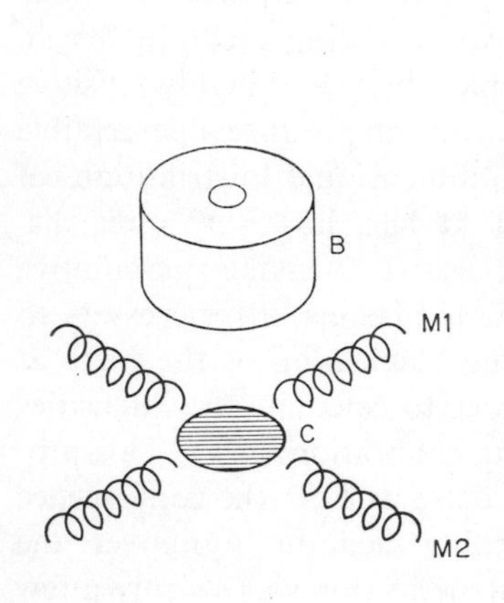

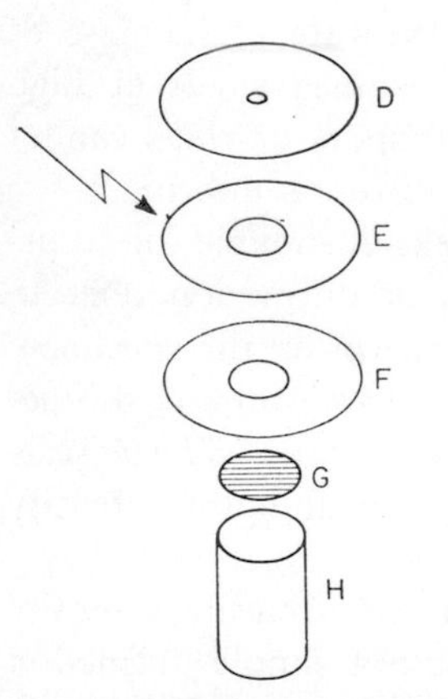

Fig. 2.16. Schematic diagram of scanning electron diffractometer.

A survey of the design requirements of a scanning diffractometer has been given by Grigson and Tillett [28]. The principal features of the instrument are illustrated schematically in Fig. 2.16. The electron beam from the source A passes through the magnetic lens B and the specimen C. It is then focused on the screen D, in which is a small aperture. Below D are two further electrodes E and F, with apertures as indicated. Electrode F is at earth potential, but electrode E is maintained at a negative potential approximately equal to that of the filament of the electron gun. Below F is a phosphor screen G, which is viewed by a photomultiplier cell H situated outside the vacuum system.

Any electrons scattered *elastically* by the specimen, and passing through the aperture in D, are able to pass through the electrodes E and F. They then strike the phosphor G, producing light which falls on the photomultiplier H. On the other hand, electrons entering D after being *inelastically* scattered are repelled by the electrode E, and do not reach F. The output from the photomultiplier is therefore a measure of the intensity of the beam of elastically scattered electrons falling on the aperture in D. A system of deflector magnets M_1, M_2 deflects the electrons leaving the specimen so that any portion of the diffraction pattern may be brought on to the aperture in D, and its intensity thus measured.

The simplest way of using the instrument is to scan along a diameter of the diffraction pattern by passing a current with saw-tooth wave-form through one of the magnets M_1 or M_2. The current through the deflector magnet also passes through a resistor, and the potential difference developed across this is applied to the X terminal of an oscilloscope. The output from the photomultiplier H is applied to the Y terminals of the oscilloscope. The resulting trace on the oscilloscope obviously represents the radial distribution of intensity in the diffraction pattern. If a more permanent record is needed, the oscilloscope is replaced by an XY pen recorder, and the scanning speed is appropriately reduced.

The two-dimensional distribution of intensity in the diffraction pattern can be exhibited by using the two magnets M_1 and M_2 to scan the pattern in a television-type raster (Grigson, Dove and Stilwell [27]. The line scan voltage is applied to the X terminals of an XY recorder. The photomultiplier output is added to the frame scan voltage, and applied to the Y terminals of the recorder. The resulting pattern, of which an example is shown in Plate 1, is a perspective view of the intensity distribution in which the intensity maxima appear as vertical spikes.

2.10 *Low energy diffraction camera*

The devices so far described operate with electron beams of energy 50–100 kev. We have noted in § 1.2 that the earliest electron diffraction experiments were carried out using low energy electrons (30–600 eV). After a long period of neglect, the technique of low energy electron diffraction (LEED for short) is now being widely used. The revival of interest in this technique is due partly to the development of improved vacuum techniques, and partly to the introduction by Scheibner, Germer and Hartman [61] of a method for directly observing a LEED pattern.

The principle of the apparatus used at the present time is illustrated in Fig. 2.17. The source of electrons is the gun E. This has the usual filament and several diaphragms to which suitable potentials can be applied so as to produce a narrow, well-collimated beam with an energy adjustable from 15 to 1000 eV. The beam from the gun travels along the metal drift tube D, which serves to screen it from external fields, and then strikes the specimen S.

The diffraction pattern is observed on the fluorescent screen P, whose shape is a hemisphere centred on the specimen S. In order that the electrons may have enough energy to excite fluorescence when they strike the screen, this is maintained at a positive potential of about 5 kV with respect to the specimen. The pattern on the fluorescent screen can be observed and photographed from outside the vacuum system through a window.

It is, of course, necessary to ensure that the trajectories of the

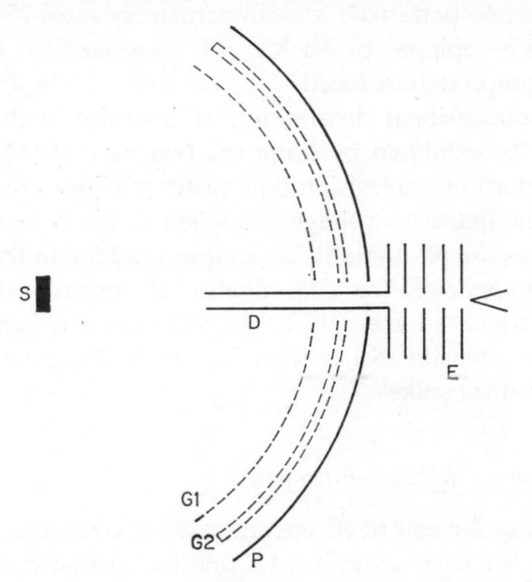

Fig. 2.17. Principle of LEED camera.

diffracted electrons are not distorted by potential gradients in the space between the specimen and the screen. For this reason, a fine wire grid G_1, of hemispherical shape, is mounted close in front of the screen. This grid is maintained at the same potential as the specimen so that, over the greater part of their journey from the specimen to the fluorescent screen, the diffracted electrons are moving in field-free space.

A further refinement is the introduction of two further hemispherical grids G_2 between G_1 and P. These grids are connected together and maintained at a potential approximately equal to that of the filament of the electron gun. Their function is to repel all electrons which have been *inelastically* scattered. The use of a *double* repeller grid was proposed by Caldwell [11], and has the advantage of producing a sharper cut-off of electrons which have lost energy.

2.11 *Low energy scanning diffractometer*

We have seen that the usual way of recording a high energy electron diffraction pattern is by means of a photographic plate, and that the scanning diffractometer was a comparatively recent innovation. In contrast, all early LEED measurements were made by direct measurement of the current in the diffracted beam. At about the same time that Germer and his associates developed the apparatus described in § 2.10 for making a LEED pattern directly visible, Farnsworth developed the alternative approach of displaying on a cathode-ray tube screen the output from an automatic scanning

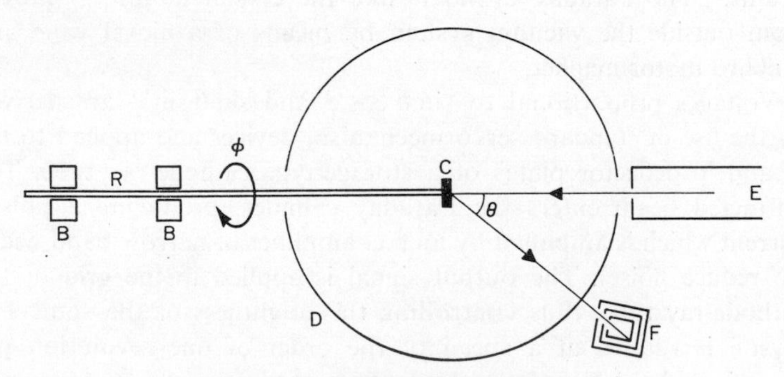

Fig. 2.18. LEED diffractometer.

diffractometer. The relative merits of the two instruments are essentially the same as those of the corresponding high energy devices: the one described in the previous section is simpler in construction, but the scanning diffractometer type is most suitable for quantitative intensity measurements.

The LEED diffractometer has been described by Farnsworth [21, 22], and Park and Farnsworth [56]. The general principle of the diffractometer is illustrated in Fig. 2.18. An electron beam from a gun E, pulsed at 10^3 p.p.s., falls normally on the crystal C. The crystal is mounted on a 3-mm diameter molybdenum rod R, which can rotate in synthetic sapphire bearings B, thus varying the azimuth ϕ of the crystal with respect to the electron beam. The rod carries nickel vanes (not shown) which enable the rod and crystal to be rotated from outside the glass-walled vacuum system by a synchro motor magnet. The nickel vane continually aligns itself with the rotating magnetic field of the synchro.

The crystal is situated at the centre of a metal drum D, which ensures that the incident and diffracted beams of electrons move in field-free space. The drum has circular apertures for the entry of the electron beam, and for the crystal mounting rod R. It has a circumferential slot (not shown) through which rays diffracted in the equatorial plane may pass. Outside the drum is a double screened Faraday cylinder F to collect the diffracted electrons. The Faraday cylinder is carried on an arm (not shown) so that it can be moved round a circle centred on the crystal, thus varying the angle θ (sometimes called the co-latitude) between the incident and diffracted beams.* The Faraday cylinder, like the crystal mount, is moved from outside the vacuum system by means of a nickel vane and synchro motor magnet.

Voltages proportional to $\sin\theta\cos\phi$ and $\sin\theta\sin\phi$ are derived by the use of standard servo mechanism devices and applied to the X and Y deflector plates of a storage-type cathode-ray tube. The diffracted beam enters the Faraday cylinder, producing a pulsed current which is amplified by an a.c. amplifier of narrow band width (to reduce noise). The output signal is applied to the grid of the cathode-ray tube, thus controlling the brightness of the spot. The crystal is rotated at a speed of the order of one revolution per second, and the Faraday cylinder is moved so as to scan through the range of the angle θ in about 90 sec. The resulting display on the cathode-ray tube screen is an orthographic projection of the sphere of reflection in reciprocal space (§ 3.3).

Besides producing a visual display of the diffraction pattern, the diffractometer can also be used to measure the intensity of the diffracted beams as a function of the energy of the electrons. To do this, an electronic circuit is used to sweep the accelerating voltage applied to the electron gun over a small range, and a signal is applied to the X plates of a cathode-ray tube which is proportional to the

* It is unfortunate that the symbol θ is conventionally used both for the Bragg angle in ordinary X-ray or electron diffraction, and for the co-latitude in LEED.

sweep voltage. The crystal is rotated so that the diffracted beam passes through the slot in the drum, and the output from the Faraday cylinder is applied to the Y plates of the cathode-ray tube. With the Faraday cylinder fixed in position, a trace such as that shown schematically in Fig. 2.19(a) is obtained. If, now, the Faraday

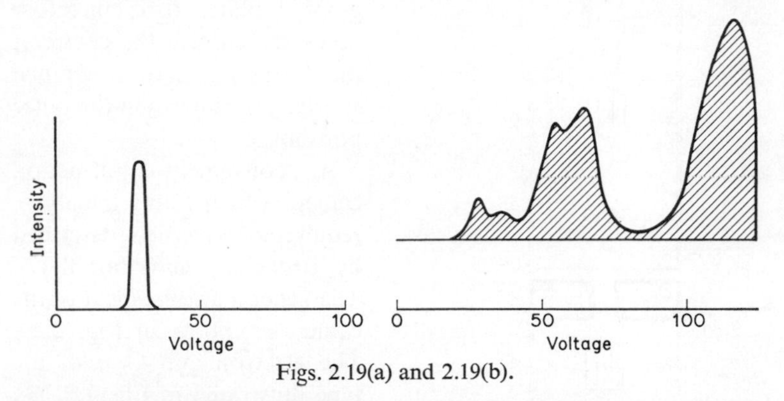

Figs. 2.19(a) and 2.19(b).

cylinder is moved slowly so as to cover a wide range of angle θ, one obtains a superposition of traces such as Fig. 2.19(a) with maxima in different positions, giving the resultant shown in Fig. 2.19(b). The envelope of this pattern is a representation of the variation of intensity of the diffracted beam with electron energy.

2.12 *Gas diffraction camera*

An important application of electron diffraction is the study of molecular structure of substances in the gas phase. The design of a camera for recording electron diffraction patterns of gases is governed by the following considerations. First, gas diffraction patterns always consist of broad haloes; consequently a high resolution type of camera is unnecessary. It is perfectly satisfactory if the diameter of the focused beam at the photographic plate is of the order of 40 μm. Secondly, there are relatively few atoms in a gaseous specimen to diffract electrons; consequently electrons which have been scattered anywhere in the camera must be prevented from reaching the photographic plate. A careful arrangement of diaphragms in the camera is needed, otherwise the stray electrons will produce a diffuse background with an intensity comparable with that of the diffraction pattern. Thirdly, there is a very great variation of intensity between the centre and the edge of the diffraction pattern. This is much greater than the extent of the linear portion of the blackening/intensity curve of a photographic plate. It is therefore necessary to give the centre of the diffraction pattern a much shorter exposure than the outer portions.

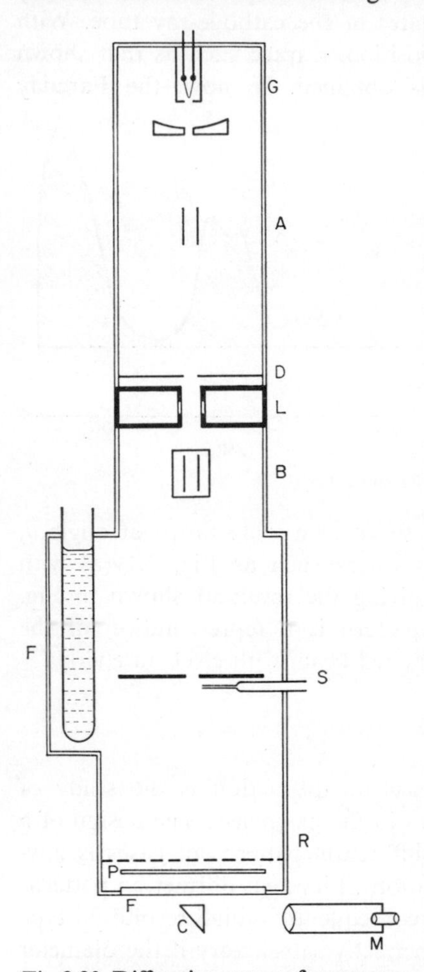

Fig. 2.20. Diffraction camera for gaseous specimens. The boundary of the camera vacuum is schematically indicated by a fine double line.

A convenient diffraction camera which fulfils the above requirements has been described by Brockway and Bartell [7]. The general arrangement of the camera is shown in Fig. 2.20. The electron gun G is of the type illustrated in Fig. 2.3. By applying a potential difference to two parallel deflector plates A situated immediately below the gun, the beam can be deflected off the diaphragm D. This is a convenient device for switching the beam on and off, and makes possible accurate electronic control of exposure.

The magnetic lens L focuses the beam on to the photographic plate P. The position of the beam can be determined when the photographic plate is removed, so that the beam strikes the fluorescent screen F. The fluorescent spot can be observed with the microscope M by reflection in the prism C. The beam can be directed to the exact centre of the fluorescent screen by means of two pairs of electrostatic plates B.

The gas sample is injected at S through a nozzle situated immediately below a diaphragm. The general arrangement of the diaphragm and nozzle, which are rigidly connected to form a single unit, is indicated in more detail in Fig. 2.21. This unit is mounted on a specimen holder by means of which it can be moved in two perpendicular directions normal to the incident beam. When the unit is positioned so that the beam passes through the diaphragm, the tip of the nozzle is only 300 μm from the beam, thus ensuring that there is a highly localized concentration of gas over a very short length of beam path.

Opposite the nozzle, a 'cold finger' F – a closed metal tube containing liquid nitrogen or similar refrigerent – protrudes into the

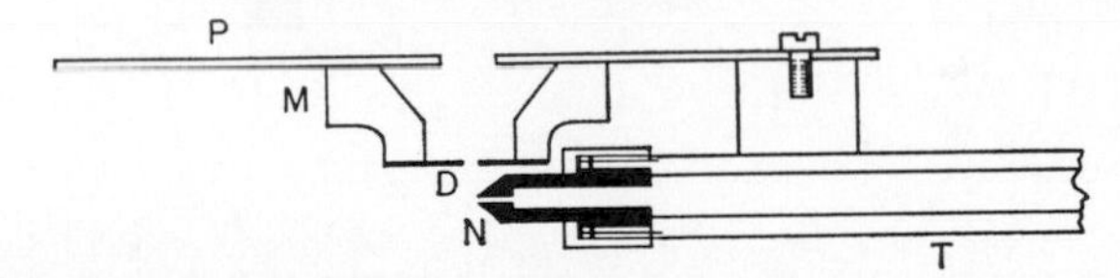

Fig. 2.21. Details of gas injection system. The nozzle N is made of platinum (for easy cleaning) and is attached to the inlet tube T by a collar. The diaphragm D is also of platinum, and is attached to a metal mount M, which in turn is attached to the plate P. The tube T is also bolted to P. The tube passes through the vacuum wall via a metal sylphon. The whole rigid assembly of nozzle, tube and diaphragm can thus be moved in the camera by means of two screws in the manner indicated in Fig. 2.6.

vacuum space of the camera. The gas from the nozzle immediately condenses on F and so does not appreciably impair the quality of the vacuum. Maintenance of a good vacuum is further assisted by having a cam-operated stopcock to control the flow of gas from the nozzle. The cam is linked with the electronic beam switch so that gas is admitted only during an actual exposure.

We have noted above that it is necessary to give a greater exposure to the outer part of the diffraction pattern. For this purpose, a rotating sector diaphragm is used. This is illustrated in Fig. 2.22, and its position in the camera is indicated in Fig. 2.20 at R. The sector device consists of a large ball-bearing having an internal diameter of about 100 mm, the outer and inner portions of the bearing being indicated by A and B respectively in Fig. 2.22. The outer part of the bearing is fixed, and a heart-shaped metal plate C is attached by a metal strip D to the rotating inner part of the bearing.

The side view section of the device shown at (b) illustrates how the portion BCD is rotated by a friction drive from a wheel W bearing on a flange F mounted on B. The shape of the sector plate C is

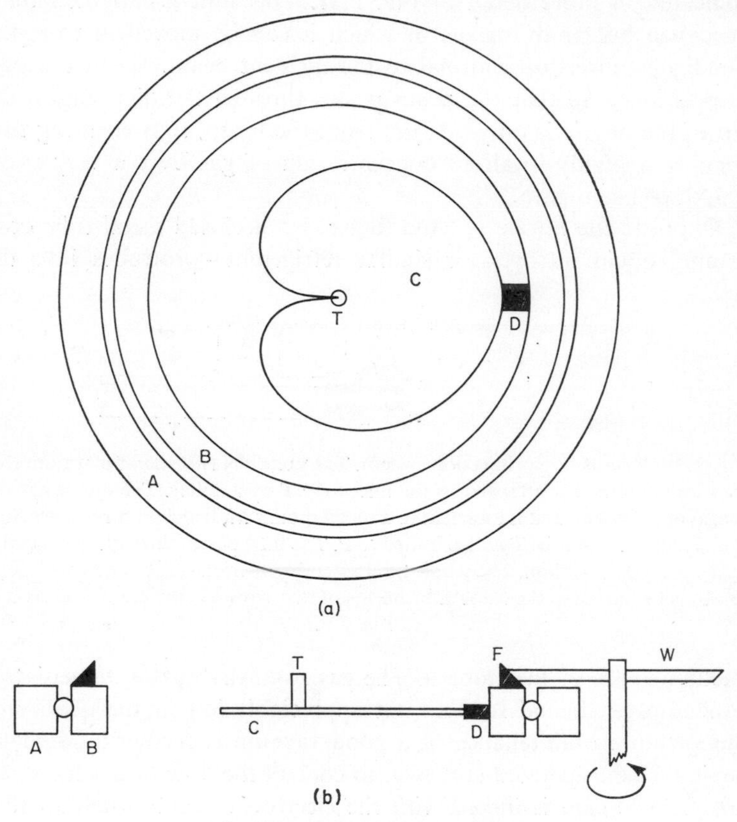

Fig. 2.22. Rotating sector device.
(a) Plan.
(b) Side view (section).
The drive system W and F seen in (b) is not shown in (a).

described in polar coordinates referred to the cusp as origin by the equation

$$r^3 = a\theta.$$

The sector plate is arranged with its cusp on the axis of rotation, and the undiffracted electron beam is directed to strike the plate at this point. The sector is rotated at a speed of about 1200 rev/min. It will easily be seen that this results in any part of the diffraction pattern being given an exposure proportional to the cube of its distance from the centre.

It is very necessary to prevent the undiffracted electron beam from reaching the photographic plate. Failure to do this produces a serious blackening due to X-rays and scattered electrons. The narrow metal tube T shown in Fig. 2.22(b) acts as an effective beam trap.

2.13 *Specimen preparation techniques*

Electrons with an energy of the order of 100 keV have a high probability of being *inelastically* scattered in traversing a distance of more than a few tens of nm in a solid composed of elements of medium atomic number. Such inelastically scattered electrons will produce a diffuse background intensity which may completely mask the diffraction pattern produced by the *elastically* scattered electrons. The first objective of any specimen preparation technique is therefore to ensure that the geometry of the specimen is such that most of the electron beam traverses only a thin region of the specimen.

Let us first consider the preparation of *reflection* specimens. It has been explained in § 1.3 and Fig. 1.5(c) that diffraction patterns from such specimens are usually produced by electrons which have passed through surface asperities. If inelastic scattering is to be unimportant, the thickness of these asperities (measured parallel to the mean surface plane of the specimen) must not exceed a few tens of nm.

Now it is clear that the grazing angle between the incident electron beam and the mean surface plane of the specimen should not exceed the smallest angle between the incident and diffracted beams, otherwise some of the diffracted beams will be cut off by the specimen. This minimum angle is typically about 10^{-2} rad. It follows that the incident electron beam will cast a shadow of any surface asperity, the length of the shadow being about 100 times the height of the asperity. The ideal surface texture for a reflection specimen therefore has asperities about 10 nm thick spaced at a distance apart equal to about 100 times their height. If the asperities are further apart than this, their number will be too few to give the most intense diffraction pattern; if they are closer, they will shadow one another from the incident beam, and again intensity will be lost.

An electron beam incident at a glancing angle of about 10^{-2} rad will impinge on a plane specimen surface over a distance of about 10 mm measured in the direction of the beam. It is therefore desirable that the specimen should be macroscopically flat over an area of about 100 mm^2, as well as having the microscopic texture described above.

The usual way of preparing a block of material is to grind the surface flat, using progressively finer grades of abrasive, and then to polish. After all traces of grease have been removed in a bath of acetone or other solvent, the surface is lightly etched in a suitable chemical reagent. To obtain the optimum surface texture requires a careful choice of reagents. This is largely a matter of trial and error. References to suitable recipes are given by Hirsch *et al.* [31]. An alternative to chemical etching is ion bombardment of the surface in the camera. An ion gun for this purpose is described in § 2.3 and illustrated in Fig. 2.9.

In the case of metals, the polishing operation may produce a polish layer of amorphous material. In such cases, the technique of electrolytic polishing is to be preferred. This consists in making the specimen the anode in a suitable electrolytic cell. With a suitable electrolyte and current density, protuberances in the anode surface are preferentially eroded, so that an initially rough surface becomes progressively smoother.

For the production of *transmission* diffraction patterns, the specimen must obviously be in the form of a film about 10 nm thick. Mica, and similar layer-type crystals, can be cleaved to produce extremely uniform films of this thickness and with an area of many mm^2; such films can be mounted directly in the diffraction camera.

Many electrically conducting materials can be prepared as thin foils for electron diffraction examination by electropolishing. The essentials of the technique are shown in Fig. 2.23. The specimen S is in the form of a disc 3 mm diameter and about 0·5 mm thick. It is mounted in a Teflon insulator T, and electrical contact is made by a fine platinum wire P. The specimen assembly is made the anode in a bath of a suitable electrolyte; a stainless steel sheet C, whose position is not critical, forms the cathode. The centre of the specimen, where it is not protected by the Teflon mount, is rapidly eroded by electrolytic action. The process is stopped when a small perforation appears in the specimen. (The use of the lamp L and microscope

M makes it easier to detect the moment when perforation occurs.) The specimen can then be removed from the Teflon mount and inserted in the diffraction camera. A region of the specimen adjacent to the perforation is likely to be the correct thickness for electron diffraction work; at the same time, the thick periphery of the specimen makes it relatively robust. Details of this, and related techniques, are given by Kay [35].

However, many materials cannot be prepared as thin films strong enough to be self-supporting, and it is necessary to mount them on a metallic, fine-mesh grid. Thin circular copper grids with a diameter

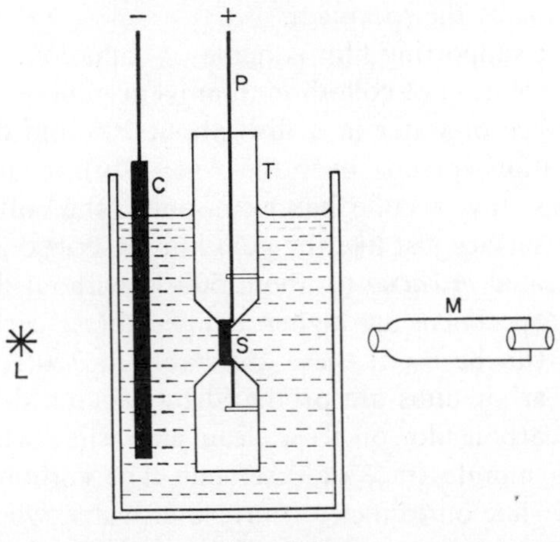

Fig. 2.23.

of 3 mm and with a mesh size of about 125 μm are available commercially as mounts for electron microscope specimens. These grids are prepared by an electrolytic process. They are cheap and, because they are usually used once only, require no cleaning.

Some specimens can be prepared as films strong enough to be mounted on such a grid without further support. For example, gold specimens can conveniently be prepared by floating a small fragment of gold leaf on the surface of a dilute solution of potassium cyanide, which slowly dissolves the metal. A specimen support grid, held in a pair of tweezers, is slipped underneath the gold film. Just before the film disintegrates, it is lifted out of the solution on the grid, transferred for a moment to a dish containing distilled water, and finally lifted out of the water on the grid and mounted in the camera, where it dries as the camera is evacuated.

Many materials cannot be prepared as films strong enough to withstand the treatment described above. It is then necessary to coat the specimen support grid with a very thin, continuous film on which particles of the specimen material can be supported. It is obviously desirable that a support film for the specimen should be strong, yet have a minimum scattering power for electrons. We shall see in Chapter 3 (equation 3.2) that this requires that it be composed of elements of low atomic number. The film must also be amorphous, so that it does not produce a diffraction pattern which would confuse that formed by the specimen.

The simplest supporting film is made of collodion. A single small drop of a 2% solution of collodion in amyl acetate is allowed to fall on to the surface of water in a dish about 200 mm diameter. The collodion solution spreads over the water surface as a very thin film. After the amyl acetate has evaporated, the collodion film is picked off the surface just like the gold film described above. Such a film can be heated *in vacuo* to about 500°K without disintegrating. For studying specimens at higher temperatures, carbon or silica support films can be used; these are stable to 750°K and 1200°K respectively. Carbon films are produced by vacuum deposition of a 10 nm thick carbon film on to a clean glass slide which has been coated with a minute trace of detergent. The carbon film is then floated off the slide on to a clean surface of water, whence it can be picked up on a grid in the manner of the gold film described above. Silica films are formed by vacuum deposition of the material on to a heat-polished block of polystyrene; the silica film is then floated off the polystyrene by immersion in an ethyl bromide–benzene mixture.

The material to be examined can be deposited on to the support film in various ways. If the substance is soluble in water, a drop of the dilute aqueous solution can be placed on the film and the water allowed to evaporate. Specimens of clay-like material can often be prepared by allowing an aqueous suspension of the clay particles to sediment on to a film-coated specimen grid. Fine powders can sometimes be dispersed as an aerosol and allowed to settle on to a coated grid. A variant of this method makes use of thermal precipitation. The aerosol is drawn down a vertical, rectangular channel about ½ mm wide, in which is mounted a horizontal, electrically heated wire. The specimen grid forms part of one wall of the channel. The aerosol particles are precipitated on to the specimen grid in the same way that air-borne dust is precipitated on to the walls of a room above a radiator. However, the most generally convenient method of specimen preparation, in the case of simpler chemical substances, is by vacuum deposition on to a coated grid. Details of all these techniques have been given by Drummond [20] and Kay [35].

It is often desirable to prepare specimens of some materials – especially metals – in the form of *single crystals* having an area of a few mm^2 and a thickness of some 10 nm. This is achieved by vacuum deposition of the material on to the freshly cleaved surface of rock salt. Rock salt is an ionic crystal, so the cleaved surface contains a regular arrangement of positively charged alkali and negatively charged halogen atoms. These produce a periodic electrostatic potential field immediately in front of the surface. Metal atoms arriving at such a surface arrange themselves in a regular array: in other words, they form a single-crystal film on the rock salt surface. (This phenomenon of the crystallographic alignment of a deposit on a substrate is known as *epitaxy*.) After the required thickness of metal has been deposited, the rock salt is removed from the vacuum and placed in warm water. This causes the metal film to float off, and it can then be picked up either on a plain specimen grid, or on one coated with a collodion or carbon film.

3

Kinematic Theory of Electron Diffraction

3.1 *Introduction*

In this chapter, we consider the diffraction of electrons by single-crystal and polycrystalline specimens. The so-called 'kinematic' theory which we employ gives a satisfactory account of the main features of the diffraction patterns obtained with a simple camera such as that described in Chapter 2. In this theory, it is assumed that only a negligible fraction of an incident electron beam is scattered by a crystal. This means that we can assume that every atom in the crystal receives an incident wave of the same amplitude: we do not have to take account of any attenuation of the incident beam as it passes through successive layers of atoms in the crystal. The total amplitude of the electron wave scattered in any direction can therefore be found by adding, with due regard to phase differences, the waves of equal amplitude scattered by every atom of the same kind in the crystal. A more refined theory must obviously take account of the fact that some electrons are scattered out of the incident beam by the first layer of atoms which it encounters. Since, therefore, the amplitude of the wave falling on the second layer of atoms is reduced, the amplitude of the wave scattered by the second layer must be less than that scattered by the first, and so on. On the other hand, the diffracted beam on its way through the crystal will lose some of its intensity by electrons being scattered back into the direction of the primary beam, which is thereby strengthened. The 'dynamical' theory, which attempts to allow for such multiple scattering effects, is discussed in Chapter 6.

It is clear that the kinematic theory will be most satisfactory when the intensities of the diffracted beams are low. These intensities diminish as P is increased, and are greater for crystals containing atoms of high atomic number. Broadly speaking, the kinematic theory is suitable for all materials if the accelerating potential is a few hundred kilovolts. For accelerating potentials in the range 10–100 kV, the theory is satisfactory for materials of low atomic number. In the case of low energy electron diffraction, only the gross features of the pattern can easily be interpreted.

3.2 *Scattering of electrons by a single atom*

Let us consider the scattering of a plane electron wave by a single atom. If the incident wave is travelling along the z axis, it may be represented by the wave function e^{ikz} where $k = 2\pi/\lambda$. Then the atom will scatter a spherical wave, which can be represented by the wave function

$$-\frac{e^{i\mathbf{k}\cdot\mathbf{r}}}{r} f(\sin\theta/\lambda), \tag{3.1}$$

where 2θ is the angle between the direction of propagation of the incident wave and the direction of the vector $\mathbf{r}$ drawn from the nucleus of the atom to a point on the scattered wave front. It is convenient to use 2θ as the angle between the incident and scattered wave directions, since we have already used θ as the Bragg angle (c.f. p. 7). It is shown in various textbooks* that

$$f(\sin\theta/\lambda) = \frac{me^2}{8h^2\pi\varepsilon_0}\frac{\lambda^2(Z-F)}{\sin^2\theta}, \tag{3.2}$$

where Z is the atomic number and $F(\sin\theta/\lambda)$ is the atomic scattering factor used in X-ray crystallography. Rationalized units are used, and ε_0 is the permittivity of vacuum. f has the dimensions of length, and is of the order of 100 pm. The function F has been tabulated for many atoms.† Making use of 1.3 and 1.4, equation 3.2 may be written

$$f(\sin\theta/\lambda) = \frac{e(Z-F)}{16\pi\varepsilon_0 P^* \sin^2\theta}. \tag{3.3}$$

3.3 *Diffraction by a crystal lattice*

In this section, we consider diffraction by a crystal which is so large that we may without serious error regard it as infinite. The effect of

* E.g. MOTT and MASSEY, *Theory of Atomic Collisions*, p. 86, Clarendon Press, Oxford.

† E.g. Ibers and Vainshtein [33].

the finite size of the crystal on the diffraction pattern is discussed in a later section.

A crystal is conveniently described by a *space lattice*. This is defined by three non-coplanar vectors $\mathbf{a}_1$, $\mathbf{a}_2$, $\mathbf{a}_3$, and the parallelepiped outlined by these vectors is the *unit cell* of the lattice. The

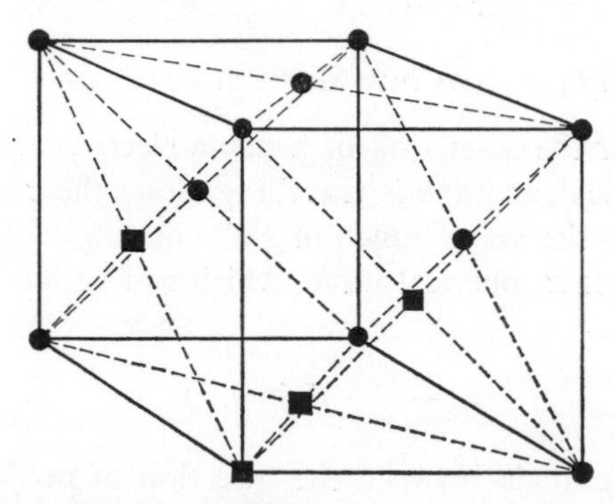

Fig. 3.1. Unit cell of face-centred cubic lattice. The points indicated by squares are the four equivalent points defining the unit cell. The remaining points, indicated by circles, are points in adjacent unit cells.

vector **R** defining the position of an atom of the crystal may be written

$$\mathbf{R} = (m_1 + i_1)\mathbf{a}_1 + (m_2 + i_2)\mathbf{a}_2 + (m_3 + i_3)\mathbf{a}_3, \qquad 3.4$$

where m_1, m_2, m_3 are integers and i_1, i_2, i_3 are proper fractions. The fundamental property of a crystal, as of a space lattice, is that it is a

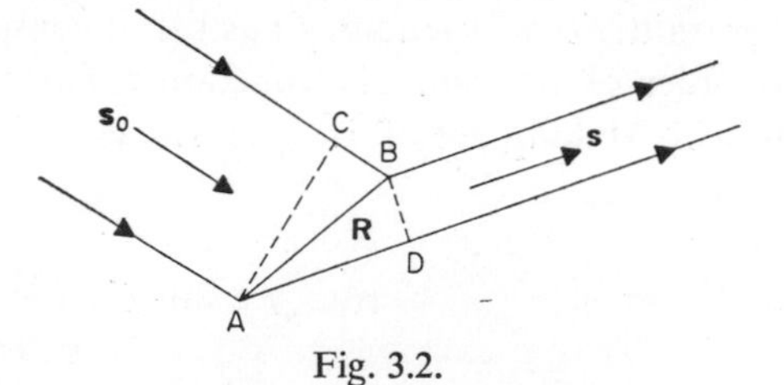

Fig. 3.2.

repeating structure. This means that if a particular atom is situated at a point given by a certain combination of numbers m and i, then an identical atom is situated at *every* point having the same values of i and every possible combination of integers m_1, m_2, m_3. A unit cell of the crystal is identified by a particular triplet of integers m. It will contain μ atoms, not necessarily of the same kind, situated at points distinguished by different values of i. For example, many metals crystallize on a face-centred cubic lattice (Fig. 3.1) the unit cell of which contains four identical atoms at points $i_1, i_2, i_3, = 0\,0\,0$; $0\,\frac{1}{2}\,\frac{1}{2}$; $\frac{1}{2}\,0\,\frac{1}{2}$; $\frac{1}{2}\,\frac{1}{2}\,0$.

In Fig. 3.2, let A be an atom situated at the origin and B be any other atom in the lattice whose position vector with respect to A is $\mathbf{R}$. Let $\mathbf{s}_0$ and $\mathbf{s}$ be *unit vectors* in the direction of the incident and diffracted beams. Then the path difference between the waves scattered by the atoms A and B is

$$\text{CB} - \text{AD} = \mathbf{R}\cdot\mathbf{s}_0 - \mathbf{R}\cdot\mathbf{s}.$$

The phase difference between the two scattered waves is therefore

$$\phi = -\frac{2\pi}{\lambda}\mathbf{R}\cdot(\mathbf{s} - \mathbf{s}_0). \qquad 3.5$$

This phase difference is most conveniently expressed in terms of the *reciprocal lattice*. The reciprocal lattice corresponding to the space lattice specified by three vectors $\mathbf{a}_1$, $\mathbf{a}_2$, $\mathbf{a}_3$ is a lattice specified by three vectors $\mathbf{b}_1$, $\mathbf{b}_2$, $\mathbf{b}_3$ such that

$$\begin{aligned} \mathbf{a}_i\cdot\mathbf{b}_j &= 0 \quad (i \neq j) \\ &= 1 \quad (i = j). \end{aligned} \qquad 3.6$$

For a cubic lattice, $\mathbf{a}_1$, $\mathbf{a}_2$, $\mathbf{a}_3$ are three equal, mutually perpendicular vectors. The reciprocal lattice vectors $\mathbf{b}_1$, $\mathbf{b}_2$, $\mathbf{b}_3$ then coincide in direction with $\mathbf{a}_1$, $\mathbf{a}_2$, $\mathbf{a}_3$ respectively, and their magnitude is the reciprocal of the magnitude of these three vectors.

In the more general case represented in Fig. 3.3, the vector $\mathbf{b}_1$ is perpendicular to the planes ABCD and EFGH (since

$$\mathbf{b}_1\cdot\mathbf{a}_2 = \mathbf{b}_1\cdot\mathbf{a}_3 = 0)$$

and its magnitude is the reciprocal of the perpendicular distance between these planes. Corresponding relations hold for the other vectors $\mathbf{b}_2$ and $\mathbf{b}_3$ defining the reciprocal lattice. Since any three pairs of parallel planes running through points of the space lattice will form a parallelepiped, the relation we have just given for the parallelepiped outlined by the vectors $\mathbf{a}_1$, $\mathbf{a}_2$, $\mathbf{a}_3$ is a particular example of a more general one. Quite generally, corresponding to any set of parallel planes running through points of the space lattice, we can draw a vector whose direction is the normal to these planes and

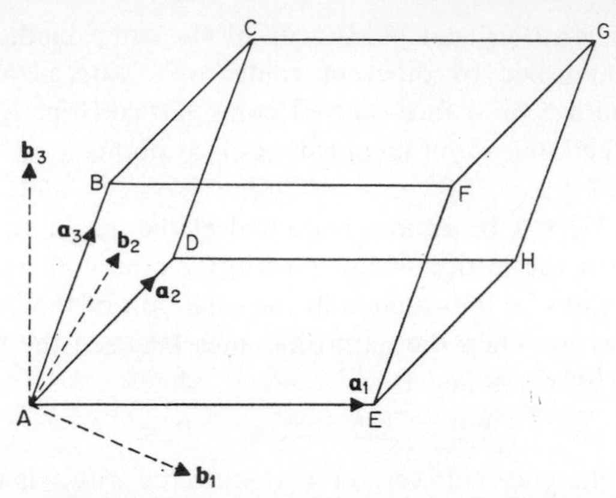

Fig. 3.3. Space lattice and reciprocal lattice vectors. The three vectors $\mathbf{a}_1$, $\mathbf{a}_2$, $\mathbf{a}_3$ define the unit cell ABCDEFGH of the space lattice. The three vectors $\mathbf{b}_1$, $\mathbf{b}_2$, $\mathbf{b}_3$ defining the reciprocal lattice are shown by broken lines.

whose magnitude is the reciprocal of their separation. The end points of such vectors constitute the reciprocal lattice.

Any vector can, of course, be expressed as a linear combination of the three vectors $\mathbf{b}_1$, $\mathbf{b}_2$, $\mathbf{b}_3$. Let us therefore write

$$\mathbf{g} = \frac{\mathbf{s} - \mathbf{s}_0}{\lambda} = g_1\mathbf{b}_1 + g_2\mathbf{b}_2 + g_3\mathbf{b}_3, \qquad 3.7$$

where the g are three scalars. Substituting 3.7 in 3.5, and making use of the relations 3.6, we obtain

$$\phi = -2\pi[(m_1 + i_1)g_1 + (m_2 + i_2)g_2 + (m_3 + i_3)g_3]. \qquad 3.8$$

From 3.1 and 3.8 it follows that at a *great distance* $\mathbf{r}$ from the origin, the wave scattered from B is

$$-\frac{e^{ik\cdot\mathbf{r}}}{r} f e^{i\phi}.$$

Hence the total wave scattered by all the atoms in the crystal is

$$\psi = -\frac{e^{ik\cdot\mathbf{r}}}{r} \sum_m \sum_i f_i$$
$$\exp\{-2\pi i[(m_1+i_1)g_1+(m_2+i_2)g_2+(m_3+i_3)g_3]\}, \qquad 3.9$$

where f_i is the quantity f of the preceding section calculated for the ith atom of the unit cell. The i summation is over all the μ atoms in a cell, and the m summation is over all the cells in the crystal. We can re-write 3.9 as

$$\psi = -\frac{e^{ik\cdot\mathbf{r}}}{r}EG, \qquad 3.10a$$

where $$E = \sum_i f_i \exp\{-2\pi i(i_1 g_1 + i_2 g_2 + i_3 g_3)\} \qquad 3.10b$$

and $$G = \sum_m \exp\{-2\pi i(m_1 g_1 + m_2 g_2 + m_3 g_3)\}. \qquad 3.10c$$

E is the *structure factor* depending on the distribution of matter in the unit cell; G is the *lattice factor*, and depends on the type of lattice and (for crystals of finite size) on the shape and size of the crystal. The intensity of the scattered wave is

$$\psi\psi^* = \frac{1}{r^2}|E|^2\,|G|^2. \qquad 3.11$$

Now

$$|G|^2 = \sum_m \sum_{m'} \exp\{2\pi i[(m_1-m_1')g_1+(m_2-m_2')g_2+(m_3-m_3')g_3]\}. \qquad 3.12$$

This has a sharp maximum when each term has the same phase, i.e. when all the g are integers. Equation 3.7 shows that $(\mathbf{s} - \mathbf{s}_0)/\lambda$ is then the position vector $\mathbf{g}$ of a point of the reciprocal lattice.

This result has been given a simple geometrical interpretation by Ewald.

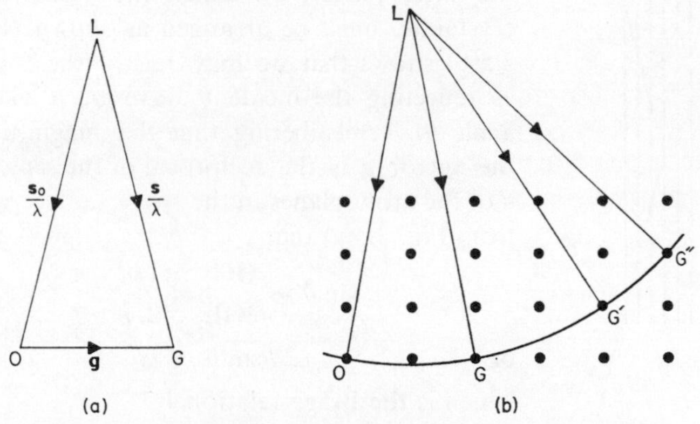

Figs. 3.4(a) and 3.4(b). Ewald reciprocal lattice construction.

In Fig. 3.4(a), O is the origin of the reciprocal lattice and G is the reciprocal lattice point whose position vector is

$$\mathbf{g} = g_1\mathbf{b}_1 + g_2\mathbf{b}_2 + g_3\mathbf{b}_3$$

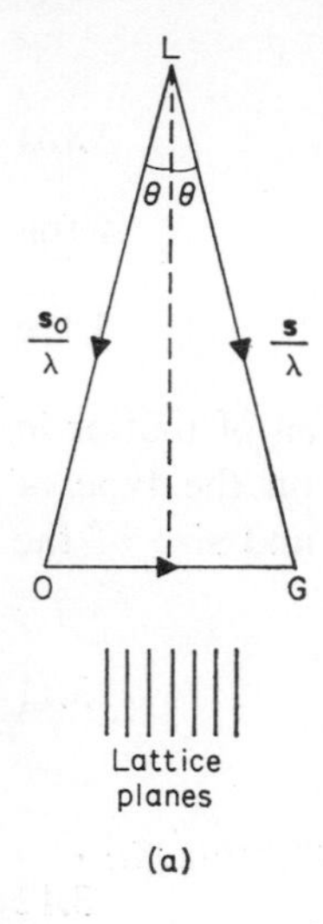

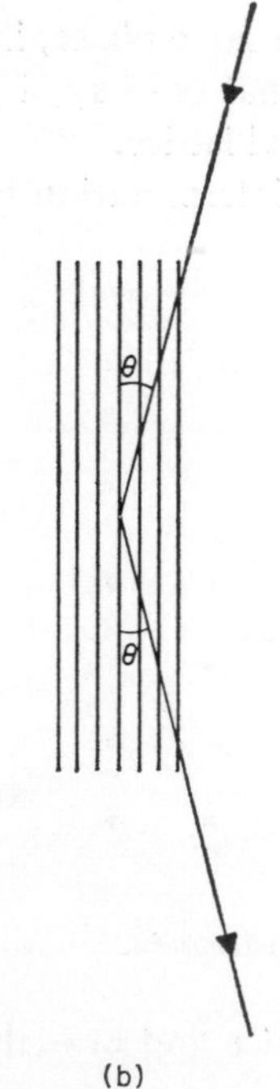

Figs. 3.5(a) and 3.5(b).

The relation 3.7 between the three vectors $\mathbf{s}_0/\lambda$, $\mathbf{s}/\lambda$ and $\mathbf{g}$ is then as shown in the figure. Remembering that $\mathbf{s}_0$ and $\mathbf{s}$ are *unit* vectors, we see that a sphere with centre at L and radius $1/\lambda$ will pass through O and G. The construction for finding the directions of the diffracted beams is therefore as shown in Fig. 3.4(b). Draw OL from the origin in the direction opposite to that of the incident beam, and of length $1/\lambda$. With L as centre, describe a sphere passing through O. If the sphere passes through any lattice points such as G, G′, G″ then LG, LG′, LG″ are the directions of the diffracted beams. The point L is usually referred to as the 'Laue point', and the sphere as the 'sphere of reflection'.

We can easily show that the reciprocal lattice construction is equivalent to the Bragg equation 1.7. Figure 3.5(a) shows the same relation between the vectors $\mathbf{s}_0/\lambda$, $\mathbf{s}/\lambda$ and $\mathbf{g}$ as Fig. 3.4(a). Since the vector $\mathbf{g}$ of the reciprocal lattice is perpendicular to a set of planes of the space lattice, the latter planes, on which the atoms of the crystal lie, must be arranged as shown. Figure 3.5(b) shows that we may regard these planes as reflecting the incident wave at a glancing angle θ. Remembering that the magnitude of the vector $\mathbf{g}$ is the reciprocal of the separation d of the atom planes in the space lattice, we see from Fig. 3.5(a) that

$$\sin\theta = \frac{\text{OG}}{2\text{OL}} = \frac{1/d}{2/\lambda}$$

or

$$2d\sin\theta = \lambda,$$

which is the Bragg relation 1.7.

In the usual high voltage electron diffraction technique, the wave-length λ is about 1/100 of the average spacing between atom planes. Hence in Fig. 3.4(b) the radius of the sphere is about 100 times as great as the distances between reciprocal lattice points. In other words, the relevant portion of the sphere of reflection is almost a plane. The importance of this is shown in Fig. 3.6, which represents the portion of a diffraction camera between the specimen at L′ and the photographic plate at O′G′. The direct, undiffracted beam strikes the photographic plate at O′ and a diffracted beam at G′. We may now superpose the construction of Fig. 3.4(b) on this, making the point L of Fig. 3.4(b) coincide with the specimen and O, G lie on L′O′ and L′G′ respectively. It is at once apparent that the pattern on the photographic plate is a projection of the section of the reciprocal lattice by the sphere of reflection. Since the portion of the sphere of reflection in which we are interested is nearly plane, we can, without serious error, regard the pattern on the photographic plate as a plane section of the reciprocal lattice. The scale factor

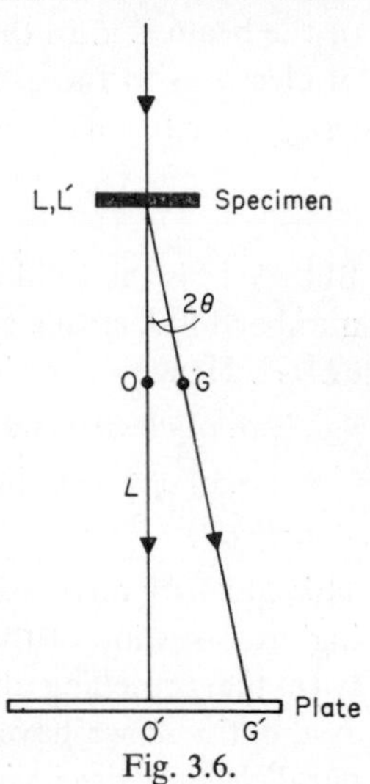

Fig. 3.6.

$$\frac{\text{scale of pattern on plate}}{\text{scale of reciprocal lattice}} = \frac{\text{O}'\text{G}'}{\text{OG}} = \lambda L \qquad 3.13$$

is the camera constant. It is this simple, direct relation which makes the reciprocal lattice so useful in interpreting diffraction patterns.

If we have a polycrystalline specimen consisting of crystals with all possible orientations, the total intensity of the diffracted beam in any direction is the sum of the intensities diffracted by all the crystals. Each crystal is associated with a reciprocal lattice having a common origin, but a different orientation. The reciprocal lattice points such as G, G′ of Fig. 3.4(b) belonging to the different crystals now merge together to form concentric spheres around the origin. The diffraction pattern is then a plane section through the centre of this system of spheres, i.e. a set of concentric circles as in Fig. 1.4(b). The radii of these circles are λL times the distances of the reciprocal lattice points from the origin.

Let us now examine more closely equation 3.11, which gives the intensity of the diffracted beam. We have assumed throughout our calculation that the incident wave has unit amplitude; this means that there is an average of one electron per unit volume in the beam. If u is the velocity of the electrons, then the number of electrons which in unit time cross a unit area set perpendicular to the direction of the beam is u. In the same way, we see from 3.11 that the number of electrons in the diffracted beam which cross an area dS set perpendicular to this beam and at a distance r from the crystal is

$$u\psi\psi^* \, dS = \frac{u}{r^2} E^2 G^2 \, dS.$$

But dS/r^2 is the solid angle subtended at the crystal by dS. Thus the number of electrons scattered into unit solid angle per unit time is uE^2G^2. Hence the ratio

$$\sigma = \frac{\text{no. of electrons diffracted per unit solid angle per unit time}}{\text{no. of electrons in incident beam crossing unit area per unit time}}$$

$$= E^2G^2.$$

This quantity obviously has the dimensions of area: it is the scattering cross-section of the crystal.

In the preceding discussion, we have made a number of tacit assumptions: we have assumed that the incident beam is perfectly parallel; that the incident electrons all have the same energy, and therefore wavelength; and that the crystal is perfectly free from distortion. None of these assumptions is strictly true. In practice, the incident beam is not quite parallel: this means that $\mathbf{s}_0$ is distributed through a narrow cone. There is a small range of wavelengths in the incident beam: this means that the radius of the sphere of reflection has a range of values. Most crystals are elastically deformed, or contain rows of dislocations. This has the same effect as if there were present a number of crystals with orientations differing slightly from one another. The combined result of all these circumstances is that the end point of the vector $(\mathbf{s} - \mathbf{s}_0)/\lambda$ is not sharply defined with respect to the reciprocal lattice, but is distributed over a small volume of this lattice. In many cases, the end point of this vector is distributed over a region which is larger than the region within which the function G differs appreciably from zero. This means that the observed scattering cross-section will have a value $\bar{\sigma}$ depending on the *mean value* of G^2. We therefore write

$$\bar{\sigma} = E^2 \frac{\int_\omega G^2 \, d\omega}{\omega}, \qquad 3.14$$

where the integration is over the volume of reciprocal space occupied by the end points of the vector $(\mathbf{s} - \mathbf{s}_0)/\lambda$. Since G^2 extends over a very small region of reciprocal space, it makes little difference whether we evaluate the integral over ω or over the whole volume Ω of a unit cell of the reciprocal lattice. We therefore write

$$\bar{\sigma} = E^2 \frac{\int_\Omega G^2 \, d\omega}{\omega}.$$

Making use of 3.12

$$\begin{aligned}\int_\Omega G^2 \, d\omega &= \sum_m \sum_{m'} \iiint \exp \{2\pi i[(m_1 - m_1')g_1 + (m_2 - m_2')g_2 \\ &\qquad + (m_3 - m_3')g_3]\}(\mathbf{b}_1 \, dg_1 \times \mathbf{b}_2 \, dg_2) . \mathbf{b}_3 \, dg_3 \\ &= \Omega \sum_m \sum_{m'} \iiint \exp \{2\pi i[(m_1 - m_1')g_1 + (m_2 - m_2')g_2 \\ &\qquad + (m_3 - m_3')g_3]\} \, dg_1 dg_2 dg_3,\end{aligned}$$

since $(\mathbf{b}_1 \times \mathbf{b}_2) \cdot \mathbf{b}_3 = \Omega$. The range of integration of each of the g's is 1.

The only non-zero integrals are those for which $m_1 = m_1'$; $m_2 = m_2'$; $m_3 = m_3'$. For these the integral is 1. Hence

$$\int_\Omega G^2 \, d\omega = \Omega \sum_m 1.$$

The number of terms in this summation is the number of sets of numbers m_1, m_2, m_3 specifying the different unit cells in the crystal: it is therefore equal to the total number, N, of unit cells in the crystal. Thus

$$\bar{\sigma} = \frac{\Omega}{\omega} N E^2. \qquad 3.15$$

The quantity Ω/ω depends on the precise experimental conditions - divergence of the incident beam, ripple in the high voltage supply, nature of the specimen. These factors are discussed later. Apart from this, we see that the intensity of the diffracted beam corresponding to any point of the reciprocal lattice is determined by E^2. The quantity E^2 takes a simple form when all the atoms in the unit cell are the same kind. For example, consider the face-centred cubic structure of most metals. If we substitute in 3.10b the values of i_1, i_2, i_3 given on p. 49 we find that

$$E = f[1 + \exp\{\pi i(g_1 + g_2)\} + \exp\{\pi i(g_2 + g_3)\} + \exp\{\pi i(g_3 + g_1)\}]. \quad 3.16$$

Since the g are integers the terms are ± 1 according as the sum of the two g's in the exponent is even or odd. Two cases present themselves: (*a*) the g's are *either* all even *or* all odd; (*b*) the g's are mixed even and odd. In the first case, the last three terms in 3.16 are all $+1$ so that $E = 4f$ and the intensity of the diffracted beam is proportional

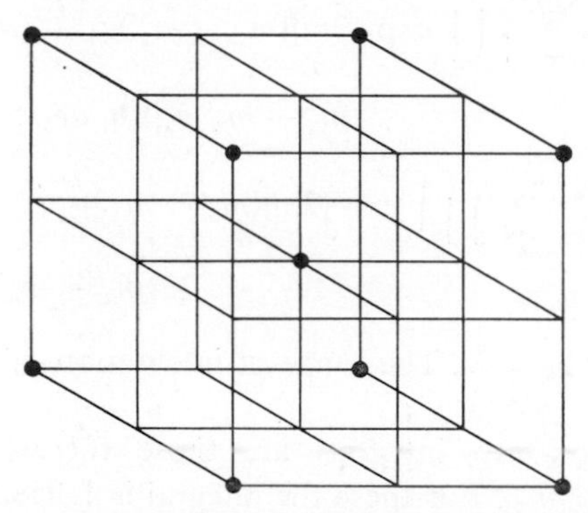

Fig. 3.7. Reciprocal lattice. The lines pass through lattice points with all integral values of g_1, g_2, g_3. Diffracted beams occur only where the sphere of reflection passes through one of the points marked with a disc: these points lie on a body-centred cubic lattice.

to $16f^2$. In the second case, two of the last three terms are -1 and one is $+1$, so that $E = 0$. Diffracted beams are therefore associated only with the points of the reciprocal lattice marked with a disc in Fig. 3.7: these points lie on a body-centred cubic lattice.

Each diffracted beam can be specified by the three integers $g_1 g_2 g_3$. These are known as the 'Laue indices' of the diffracted beam. It can be shown that the Laue indices of any diffracted beam are identical with the crystallographic Miller indices specifying the direction of the planes of atoms which produce this diffraction, or else are a common integral multiple of these. We see that, in the present case, the Laue indices are *either* all odd *or* all even. The Laue indices of the points of the reciprocal lattice of Fig. 3.7 are, in ascending order of their distance from the origin, 111, 200, 220, 311, . . .; this is also the increasing order of the distance of the diffracted beams from the centre of any diffraction pattern.

Let us now see how the general expression 3.15 can be applied to calculate the intensities of the diffracted beams. Suppose we have a polycrystalline transmission specimen giving a diffraction pattern consisting of uniform concentric rings (Debye–Scherrer pattern) as in Fig. 1.4(b). There are many crystallites in the specimen oriented at random in all possible directions with respect to the incident beam. In Fig. 3.4(b), we should have a very large number of reciprocal lattices, each with a common origin O, but oriented in all possible ways with respect to the figure OLGG′. In other words, the reciprocal lattice points would be smeared out into concentric spheres round the origin. The diffraction pattern (cf. p. 53) is a nearly plane section of this system of spheres, i.e. a system of concentric circles.

Instead of considering the reciprocal lattice to take up all possible orientations with respect to a fixed figure OLGG′, it is often more convenient to regard the lattice as fixed and OG etc. to pivot about O into all possible orientations. The end point G of the vector **g** will then describe a sphere in reciprocal space. If the incident or diffracted beams have a small angular divergence, then the magnitude of the vector g will vary over a small range Δg, and the end point G will then range over a *spherical shell* of radius g and thickness Δg. The volume of this shell is the quantity ω of 3.15. Hence

$$\omega = 4\pi g^2 \, \Delta g.$$

Making use of the relation 3.13 between the scale of the diffraction pattern and the scale of the reciprocal lattice, we see that if the diffraction ring is of radius R and width ΔR

$$\omega = 4\pi R^2 \, \Delta R/(\lambda L)^3.$$

Hence from 3.14 and 3.15, if I_0 is the flux of electrons in the incident beam (number of electrons crossing unit area per unit time), then the number of electrons diffracted into unit solid angle per unit time is

$$\frac{I_0 \Omega N E^2 (\lambda L)^3}{4\pi R^2 \, \Delta R}. \qquad 3.17$$

Now the area of the diffraction ring is $2\pi R \, \Delta R$, and its distance from the specimen is, very nearly, L. Hence the electrons giving rise to the diffraction ring must be contained in a solid angle $2\pi R\,\Delta R/L^2$. The expression 3.17 now shows that the number of electrons diffracted per unit time into the ring is

$$I_0 \Omega N E^2 \lambda^3 L/2R.$$

In terms of i_0, the current carried by the primary beam, i, the current flowing into the complete diffraction ring, and A, the area of the specimen irradiated, this may be written

$$i/i_0 = \Omega N E^2 \lambda^3 L/2RA.$$

If v is the volume of the unit cell of the crystalline material, $\Omega = 1/v$. Nv is the total volume of the specimen irradiated, and is equal to At, where t is the thickness of the specimen. Further, we may write (1.9) $\lambda L = Rd$. With these substitutions, the above expression for i/i_0 becomes

$$i/i_0 = E^2 \lambda^2\, td/2v^2. \qquad 3.18$$

Usually one would not measure the current flowing into the complete diffraction ring, but the current i' passing through a slit of length l placed tangentially to the ring. Such a slit would admit a fraction $l/2\pi R$ of the electrons diffracted into the complete ring. Hence

$$i'/i_0 = E^2 \lambda^2\, tdl/4\pi v^2 R. \qquad 3.18a$$

In principle, all the quantities in 3.18a with the exception of E can be measured, and this equation can therefore form the basis of an absolute determination of E. In practice, difficulties arise. The specimen, or its supporting grid, may absorb some of the incident beam, and then the value obtained for E will be too low. On the other hand, if i_0 is taken to be the current in the transmitted, undiffracted beam which goes into the centre of the pattern, any holes or thin patches in the specimen (and it is difficult to avoid these in the case of specimens thin enough to produce satisfactory diffraction patterns) will give too high a value for E. Most investigators have therefore been content to make relative measurements of the intensities of the diffraction rings, and to deduce from these the ratio of the structure factors $E(g_1g_2g_3)$ of diffractions of different Laue indices.

3.4 *Effect of the finite size of the diffracting crystals*

In the foregoing theory, we have assumed that a crystal diffracting the electrons contains a very large number of unit cells. The consequence of this is that the function G^2 of 3.12 is appreciably different from zero only at the points of the reciprocal lattice, where it has a sharp maximum. The discussion of p. 53 shows that the diffraction pattern of a single crystal then consists of small spots, and that of a polycrystalline specimen of a series of sharp rings, the width of the spots or rings depending only on such purely instrumental effects as fluctuations in the h.t. potential or divergence of the primary electron beam. If, however, the number of unit cells in a crystal is relatively small, so that the range of integers m, m' in 3.12 is restricted, then $|G|^2$ will be different from zero even when the g's of that equation are non-integral. This means that the function $|G|^2$ extends over a finite volume of space around each lattice point. We note that if any of the g in 3.12 are changed by an integer, the expression remains unchanged in value. Hence $|G|^2$ is a function in reciprocal space having the periodicity of the reciprocal lattice, i.e. each reciprocal lattice point is surrounded by a *region of the same size and shape* in which $|G|^2$ differs from zero. In place of the construction of Fig. 3.4(b) for finding the direction of the diffracted beam, we have that of Fig. 3.8. Here we have a perfectly collimated primary beam of monoenergetic electrons represented by the line LO incident on a single perfect crystal; but the diffracted beams are spread out over the finite cones shown. The diffraction pattern of such a finite single crystal will therefore consist of spots of finite size; and the pattern from a polycrystalline aggregate will consist of rings of finite width.

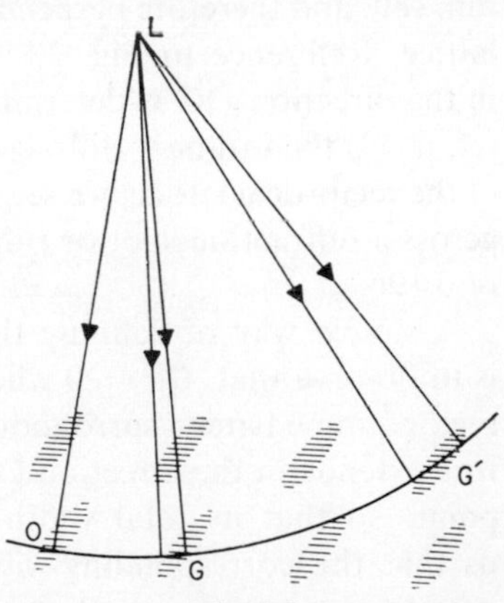

Fig. 3.8.

In order to examine the nature of this effect in more detail, let us assume that we have a crystal in the form of a parallelepiped with edges parallel to the sides $\mathbf{a}_1$, $\mathbf{a}_2$, $\mathbf{a}_3$ of the unit cell, and of length

$(N_1 - 1)a_1$, $(N_2 - 1)a_2$, $(N_3 - 1)a_3$ respectively. Then in 3.10c, m_1 can range from 1 to N_1, and similarly for m_2 and m_3. Hence

$$G = \exp\{\pi i(N_1 - 1)g_1\} \frac{\sin \pi N_1 g_1}{\sin \pi g_1} \exp\{\pi i(N_2 - 1)g_2\} \frac{\sin \pi N_2 g_2}{\sin \pi g_2} \exp\{\pi i(N_3 - 1)g_3\} \frac{\sin \pi N_3 g_3}{\sin \pi g_3} \qquad 3.19a$$

so that

$$|G|^2 = \frac{\sin^2 \pi N_1 g_1}{\sin^2 \pi g_1} \frac{\sin^2 \pi N_2 g_2}{\sin^2 \pi g_2} \frac{\sin^2 \pi N_3 g_3}{\sin^2 \pi g_3}. \qquad 3.19b$$

Let us now restrict our study still further by supposing that the reflecting planes in the crystal are parallel to the vectors $\mathbf{a}_2$, $\mathbf{a}_3$ of the unit cell, and therefore perpendicular to the vector $\mathbf{b}_1$ of the reciprocal lattice. Reference to Fig. 3.8 shows that the variation of intensity in the direction OG is determined by the first factor of 3.19b. Since (cf. p. 53) the ordinary diffraction pattern represents a plane section of the reciprocal lattice, we see that the radial distribution of intensity across a diffraction spot or ring is also determined by the first factor of 3.19b.

A simple way of defining the width of a spot or diffraction ring is to observe that $|G|^2 = 0$ when $g_1 = \text{integer} \pm 1/N_1$. The intensity region immediately surrounding a reciprocal lattice point therefore extends in the direction OG to b_1/N_1 either side of the lattice point, so that its total width is $2b_1/N_1$. Equation 3.13 then tells us that the corresponding width of a spot or ring of the diffraction pattern is

$$\frac{2b_1}{N_1}\lambda L = \frac{2}{N_1 d}\lambda L,$$

where d is the spacing of the reflecting planes, which we have assumed to be perpendicular to the vector $\mathbf{b}_1$ of the reciprocal lattice. Using 1.9, we can then express the width ΔR of a diffraction ring of radius R, or the width of a spot at a distance R from the centre of the pattern, as

$$\Delta R = 2R/N_1. \qquad 3.20$$

This simple approach, while correct in principle, is in practice not convenient. If we have a polycrystalline specimen, the sizes of the crystals will usually be distributed about a mean value. The diffraction pattern will then consist of the superposition of a large number of intensity functions of the form $\sin^2 N_1\pi g_1/\sin^2 \pi g_1$ having different values of N_1. The effect, illustrated schematically in Fig. 3.9, is to produce an intensity function which is roughly Gaussian in shape, and falls asymptotically to zero on either side of the maximum. It turns out that the most meaningful definition of the width of such a curve is the so-called *integral width* defined by

$$\text{Integral width} = \frac{\int_{-\infty}^{+\infty} I\,dR,}{I_m} \qquad 3.21$$

i.e. the width of a hypothetical rectangular intensity distribution (shown dotted in Fig. 3.9(d)) which has the same maximum and total intensity as the actual distribution.

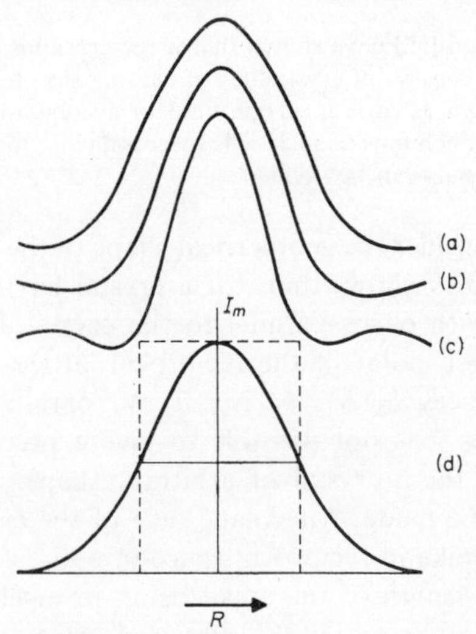

Fig. 3.9. The superposition of intensity distributions of the form of (a), (b), (c), etc. yields a distribution of the form of (d).

The integral width in reciprocal space of the intensity distribution round each lattice point is, in the case of a single crystal in the shape of a parallelepiped, given by

$$\int_{g_1=-\frac{1}{2}}^{g_1=+\frac{1}{2}} \frac{\sin^2 N_1\pi g_1}{\sin^2 \pi g_1} d(g_1 b_1) \bigg/ \lim_{g_1\to 0} \frac{\sin^2 N_1\pi g_1}{\sin^2 \pi g_1}.$$

It is justifiable to restrict the range of integration to a single cell in reciprocal space because we assume the crystal to be large enough for the extent of the function $|G|^2$ to be small compared with the unit cell. The numerator can be found from contour integration to be b_1N_1, and the denominator is N_1^2. The integral width measured in reciprocal space is therefore b_1/N_1, and the integral width of a diffraction ring of radius R, or the width of a spot distant R from the centre of the pattern is

$$\Delta R = R/N_1. \qquad 3.22a$$

It follows from equations 3.22a and 1.9 that the angular divergence of the diffracted beam is

$$\delta\varepsilon = \Delta R/L = \lambda/w, \qquad 3.22b$$

where $w = N_1d$ is the width of the specimen normal to the reflecting planes of atoms.

Stokes and Wilson [62] have shown that if the crystal is not a parallelepiped, or if the specimen consists of crystallites of varying size, the experimental value of N_1 obtained from 3.22a is a certain kind of average of the number of unit cells in the different columns of such cells (perpendicular to the reflecting planes) into which the crystals can be divided.

Let us now consider the geometrical shape of the intensity function $|G|^2$. Equation 3.19 shows that, for a crystal having the shape of a parallelepiped with edges parallel to the crystal axes, the intensity region round each point of the reciprocal lattice consists of spikes extending distances b_1/N_1, b_2/N_2, b_3/N_3 parallel to the axes in reciprocal space. It is not possible to give a precise picture of the intensity region for a crystal of arbitrary shape, but some general statements can be made. Any plane face of the crystal corresponds to an intensity spike in reciprocal space in a direction perpendicular to the face, the length of the spike being inversely proportional to the thickness of the crystal in this direction. Also, the intensity region always has a centre of symmetry at the reciprocal lattice point irrespective of whether the external form of the crystal has a centre of symmetry. Hence, crystals of both tetrahedral and octahedral form have intensity regions consisting of eight spikes.

It would be difficult to obtain a sufficiently strong diffraction pattern from just one polyhedral crystal which was sufficiently small to produce spikes of appreciable extent. However, thin metal films grown epitaxially on rock salt (p. 45) sometimes have sub-microscopic facets. Instead of a simple spot pattern, we then get either

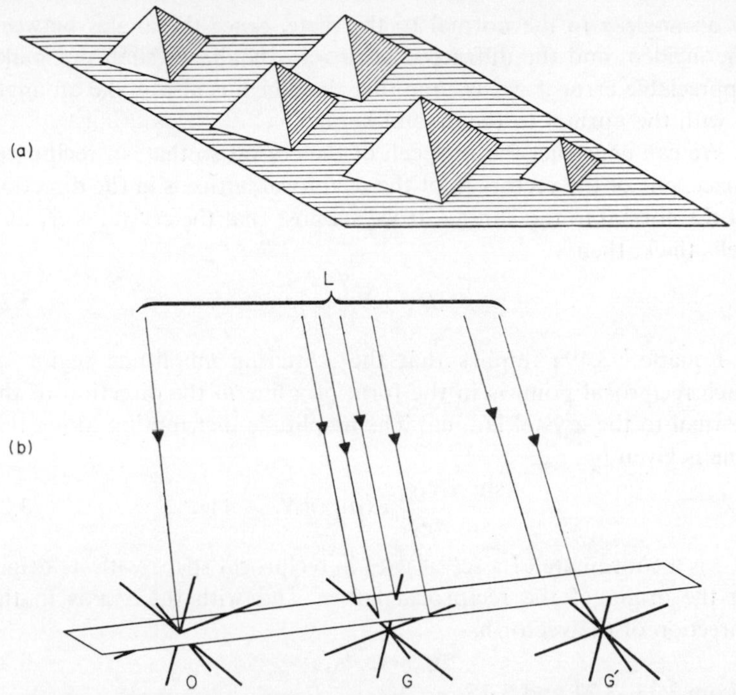

Figs. 3.10(a) and (b). If a specimen consists of a single crystal in the form of a thin film parallel to 100 planes, and having octahedral facets as indicated in (a), then the intensity regions in reciprocal space will be in the form of spikes as shown at O, G, G′ in (b). LO is the incident beam. The figure shows a portion of the sphere of reflection (approximately a plane) intersecting the intensity regions round the reciprocal lattice points G, G′ to give a group of four beams LG and two beams LG′.

short streaks intersecting at the point where a diffraction spot would be expected, or else a group of spots surrounding this point. The way in which these effects occur will be seen from Fig. 3.10. Plate 2 is an example of such streaks and spots. The study of such patterns can sometimes provide information about the external form of a crystal which it would be difficult to obtain by other means.

3.5 *Convergent beam patterns produced by a crystal lamina*

Equation 3.19 can be applied to the problem of diffraction by a crystal plate of thickness D tilted so that an incident beam is inclined at an angle α to the normal to the plate. Since the angles between the incident and the diffracted beams are small, we shall not make appreciable error if we assume that these beams also make an angle α with the normal to the crystal lamina.

We can construct the unit cell of the crystal so that, in reciprocal space, one of the vectors $\mathbf{b}_1$ of the reciprocal lattice is in the direction of the normal to the lamina. If we suppose that the crystal is N_1 unit cells thick, then

$$D = N_1\left(\frac{1}{b_1}\right). \qquad 3.23$$

Equation 3.19a implies that the scattering amplitude region at each reciprocal point is in the form of a line in the direction of the normal to the crystal lamina. The amplitude distribution along this line is given by

$$G = \frac{\sin \pi N_1 g_1}{\sin \pi g_1} \exp \{\pi i(N_1 - 1)g_1\}. \qquad 3.24$$

If z is a coordinate of a set of axes in reciprocal space with its origin at the origin of the reciprocal lattice, and with the z axis in the direction of the vector $\mathbf{b}_1$

$$g_1 = z/b_1. \qquad 3.25$$

From 3.23, 3.24 and 3.25

$$G = \frac{\sin \pi Dz}{\sin (\pi z/b_1)} e^{\pi iDz} e^{-\pi i g_1}. \qquad 3.26a$$

Because of the periodic nature of the function 3.26a, we can equally well regard z as measuring the distance of a point in reciprocal space from a lattice point. If the thickness of the crystal is greater than a few tens of unit cells, G will be appreciable only near a lattice point, so that z will be small and we can replace the sine in the denominator of 3.26a by its argument

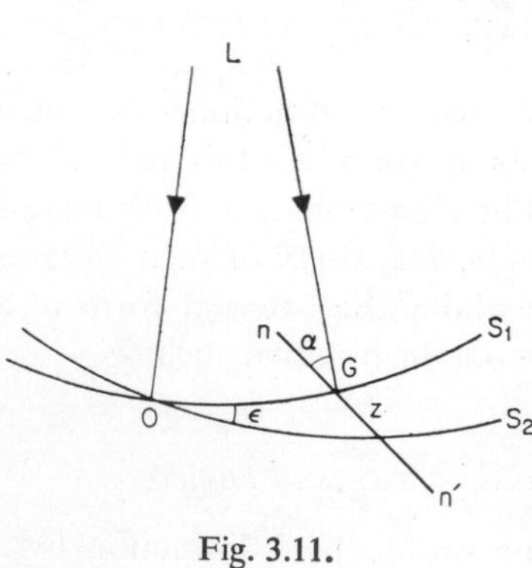

Fig. 3.11.

$$G = \frac{\sin \pi Dz}{\pi z/b_1} e^{\pi iDz} e^{-\pi i g_1} \qquad 3.26b$$

If the crystal is set at exactly the Bragg angle for a particular diffraction, then the sphere of reflection S_1 passes through the reciprocal lattice point G (Fig. 3.11). If the crystal is tilted so as to change the glancing angle with the reflecting planes by a small amount ε, then S_2 will be the new position of the sphere of reflection. This intersects the intensity spike nGn′ through G at a distance z which is seen to be

$$\begin{aligned} z &= \mathrm{OG}\varepsilon \sec \alpha \\ &= g\varepsilon \sec \alpha, \end{aligned} \tag{3.27}$$

where $g = 1/d$ is reciprocal of the interplanar spacing corresponding to the diffraction. Inserting this value of z in 3.26a

$$G = \frac{\sin (\pi \varepsilon g D \sec \alpha)}{\pi \varepsilon g \sec \alpha / b_1} e^{\pi i \varepsilon g D \sec \alpha} e^{-\pi i g_1}. \tag{3.28a}$$

The intensity of the diffracted beam is proportional to

$$|G|^2 = \frac{\sin^2 (\pi \varepsilon g D \sec \alpha)}{(\pi \varepsilon g \sec \alpha / b_1)^2}. \tag{3.28b}$$

This equation shows how the intensity of the diffracted beam varies with ε, the angle by which the primary beam deviates from the exact Bragg position.

The intensity variation shown by 3.28b is exhibited in convergent beam diffraction patterns (§ 2.7). The patch produced by the diffracted beam is seen to be crossed by parallel interference fringes. These correspond to the minima of the function $|G|^2$, and occur where

$$\varepsilon g D \sec \alpha = \text{integer}.$$

If L is the specimen-to-plate distance, the separation of these fringes is then

$$x = \frac{L}{gD \sec \alpha} = \frac{Ld}{D \sec \alpha}. \tag{3.29}$$

The thickness D of a crystal plate can therefore be determined by measuring the fringes in this type of pattern.

It is found that the fringes near the centre of the diffracted patch have abnormal separations. This is due to dynamical interference effects, and is discussed further in § 6.3.

4

The Interpretation of Electron Diffraction Patterns

4.1 *Lattice constant measurement*

The most obvious information which can be derived from an electron diffraction pattern is the determination of the crystal lattice. Let us suppose we have a transmission pattern from a polycrystalline specimen. If the crystallites are oriented completely at random, the pattern will consist of concentric rings (§ 1.3). We can measure the radii R of the rings and then, from 1.9, we can calculate the corresponding interplanar spacings d *provided we know the camera constant* λL. The usual way of determining the camera constant is to record a diffraction pattern using a substance of known crystal structure. The most commonly used standard specimen is made by vacuum deposition of thallium chloride on a thin collodion or carbon film. This material forms body-centred cubic crystals, the length of the side of the unit cell being 383·4 pm. The crystals are large enough to produce sharp diffraction rings, yet small enough for the number of them in the area of the specimen irradiated by the electron beam to be so large that the pattern consists of continuous rings rather than separate spots.

It is fairly easy with a simple diffraction camera of the type described in § 2.1 to make lattice constant measurements with an accuracy of 0·1%. However, when using an electron microscope to produce a selected area diffraction pattern (§ 2.8) it is difficult to place, with sufficient precision, the comparison thallium chloride specimen in the same position as the specimen under examination. This is because the effective camera length is critically dependent on the relative position of the specimen and objective lens. The resulting error is usually of the order of a few per cent.

If the specimen consists of cubic crystals, the crystal class is easily identified by noting the ratios between the radii of the various diffraction rings. Let us suppose, for example, that we are dealing with face-centred cubic crystals. Figure 3.7 shows that the coordinates of the reciprocal lattice points are, in units of b, the side of the unit cell in reciprocal space; 111, 200, 220, 311, . . . These points are at distances from the origin b ($\sqrt{3}$, $\sqrt{4}$, $\sqrt{8}$, $\sqrt{11}$, . . .). For a polycrystalline specimen the reciprocal space diagram consists of spheres with these radii, and the diffraction pattern is a plane section of this through the origin scaled up by a factor λL (3.13). Since, from the definition of the reciprocal lattice, $b = 1/a$ where a is the side of the unit cell, the pattern therefore consists of concentric circles of radii $(\lambda L/a)$ ($\sqrt{3}$, $\sqrt{4}$, $\sqrt{8}$, $\sqrt{11}$, $\sqrt{12}$). In the same way, we can show that a body-centred cubic specimen will give a diffraction pattern of circles of radii $\sqrt{2}(\lambda L/a)$ ($\sqrt{1}$, $\sqrt{2}$, $\sqrt{3}$, $\sqrt{4}$, $\sqrt{5}$, . . .); while for a diamond type material the radii are $(\lambda L/a)$ ($\sqrt{3}$, $\sqrt{8}$, $\sqrt{11}$, $\sqrt{16}$, $\sqrt{19}$).

EXAMPLE

An unknown material has a diffraction pattern consisting of rings of diameter 20·8, 24·0, 34·0, 39·8 mm. The diffraction pattern from a comparison thallium chloride specimen has rings of diameter 18·6, 26·3, 32·2, 37·2 mm. What is the lattice type and lattice constant of the unknown material?

Diffraction patterns obtained by 'reflection' (p. 8) can be interpreted in the same way as the transmission patterns described above. However, it is difficult to measure the diameter of the rings to better than 1% because, as is evident from Fig. 1.5, the distance L from the specimen to the photographic plate is indeterminate to the extent of the length of the position of the specimen irradiated by the electron beam: this is commonly a few mm.

4.2 *Determination of the unit cell from single-crystal patterns*

Diffraction patterns from single crystals can be obtained with a simple camera if a monocrystalline specimen can be produced whose lateral dimensions are of the order of 1 mm, but whose thickness is not more than about 100 nm. Such specimens can be produced by epitaxial growth or, in the case of hard materials, by grinding and etching a thin plate. Single-crystal patterns can also be produced from constituents of materials such as clays, by the method of selected area diffraction (§ 2.8). In either case, the crystal will probably have a low index plane parallel to the specimen plane, and the diffraction pattern will represent an important section of the reciprocal lattice.

The general principles involved in the interpretation of such single-crystal patterns can be demonstrated by reference to the idealized pattern represented in Fig. 4.1. We will examine this pattern to

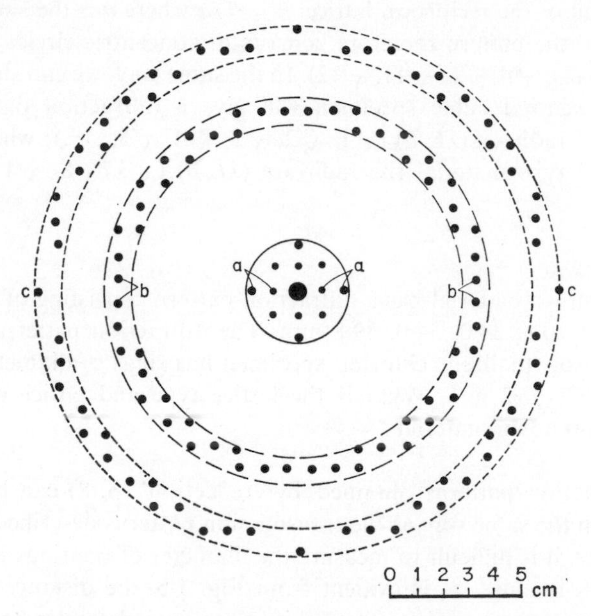

Fig. 4.1. Sketch showing diffraction pattern from a single-crystal specimen in the form of a thin lamina set perpendicular to the electron beam. $L = 500$ mm; $\lambda = 4{\cdot}5$ pm ($P^* = 67$ kV).

discover the shape and size of the unit cell of the crystal. The pattern consists basically of a square net of spots with a mesh size of 9·0 mm. Since the 'camera constant' λL is 2·25 μm^2, it follows from equation 3.13 that the section of the reciprocal lattice represented by the centre of the pattern is a square net with mesh size 4 nm^{-1}.

It will be observed that the spots in Fig. 4.1 are grouped in concentric circular zones; these are referred to as Laue zones. The origin of these zones can be explained by reference to Fig. 4.2. In this figure, O is the origin of the reciprocal lattice and L is the Laue point. LO is the direction of the incident beam. The figure shows a section of the sphere of reflection passing through O. Since the crystal is supposed to be in the form of a thin lamina perpendicular to the incident beam, the reciprocal lattice points are each surrounded by an intensity region in the form of a short spike extending a distance $1/t$ on either side of the lattice point, where t is the thickness of the crystal (§ 3.4). For the sake of clarity, the curvature of the sphere of reflection has been exaggerated; if the sphere were drawn on the same scale as the bottom row of points of the reciprocal lattice, it would be indistinguishable from a plane. To conform with the

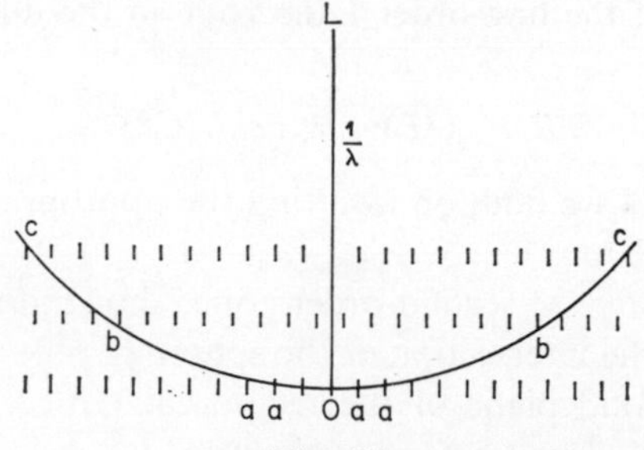

Fig. 4.2.

exaggerated curvature of the sphere of reflection, the vertical scale of the reciprocal lattice has also been increased; this also serves to make the intensity spikes of perceptible length.

The sphere of reflection will be observed to intersect intensity spikes marked *a*, *b*, *c*. The corresponding spots in the pattern of Fig. 4.1 are marked with the same letters. From the geometry of the circle, we see that the distance g_0 from the line LO of a reciprocal lattice point in the bottom row whose intensity spike is just intersected by the sphere is given by

$$g_0^2 \approx \left(\frac{1}{t}\right)\left(\frac{2}{\lambda}\right). \qquad 4.1$$

Hence, from equation 3.13, the spots at the centre of the pattern must be contained in a circle of radius

$$R_0 = (\lambda L)g_0 = (2\lambda L^2/t)^{\frac{1}{2}}. \qquad 4.2$$

In Fig. 4.1, the *zero-order Laue zone*, indicated by a circle, has a radius $R_0 = 18$ mm. Hence, from 4.1 and 4.2, and using the data set out below Fig. 4.1, we find $t = 6{\cdot}9$ nm.

The first-order Laue zone, indicated by the dashed circles in Fig. 4.1, has a mean radius of 67 mm. Figure 4.2 shows how this zone is formed by the intersection of the sphere of reflection with points of the reciprocal lattice lying in the plane adjacent to the plane through O. It is evident from the geometry of the sphere that the radius g, of the first-order zone is given by

$$g_1^2 = C^*\left(\frac{2}{\lambda}\right) \tag{4.3}$$

where C^* is the separation of the planes of the reciprocal lattice. The radius R_1 of the first-order Laue zone in the diffraction pattern is therefore

$$R_1 = (\lambda L)g_1 = (2\lambda L^2C^*)^{\frac{1}{2}} \tag{4.4}$$

From 4.3 and 4.4 we find, on inserting the appropriate values of R_1, λ and L, $C^* = 2\ \text{nm}^{-1}$.

In the same way the second-order zone, shown dotted in Fig. 4.1, corresponds to the intersection of the sphere of reflection with points lying in the second plane of the reciprocal lattice. In general, the radius R_n of the nth-order zone is given by

$$R_n = (\lambda L)g_n$$

where $$g_n^2 = nC^*\left(\frac{2}{\lambda}\right).$$

We have now found that the reciprocal lattice corresponding to Fig. 4.1 consists of parallel planes, each plane consisting of a square net of points having a mesh size of 4 nm^{-1}, the planes being separated by 2 nm^{-1}. Closer examination of Fig. 4.1 will show that the spots lying in the first-order zone do not lie on the same net as those in the zero- and second-order zones. If the net formed by the spots in the zero- and second-order zones is extended to cover the first-order zone, it will be found that the spots of the latter are each situated at the *centre* of a mesh of such a net. The explanation of this is that the reciprocal lattice is a body-centred cubic structure having a unit cell of side 4 nm^{-1}. Planes parallel to the base each contain points arranged in a square net of mesh size 4 nm^{-1}. Adjacent planes are separated by half this distance, and are displaced with

PLATES

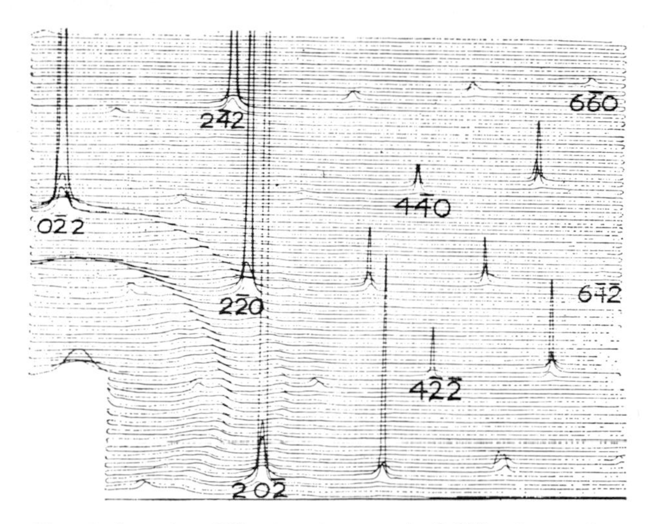

Plate 1. Scanning diffractometer record of diffraction pattern of gold film.

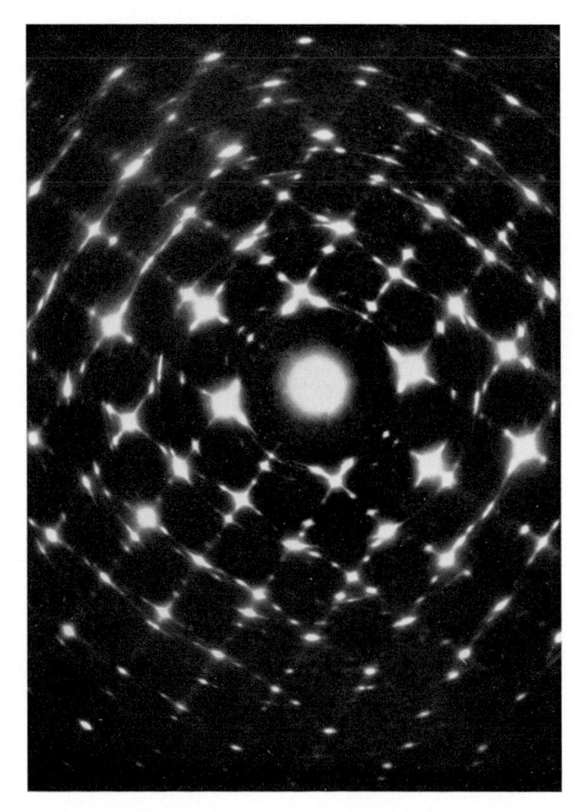

Plate 2. Single-crystal transmission pattern showing streaks and spots due to shape effects.

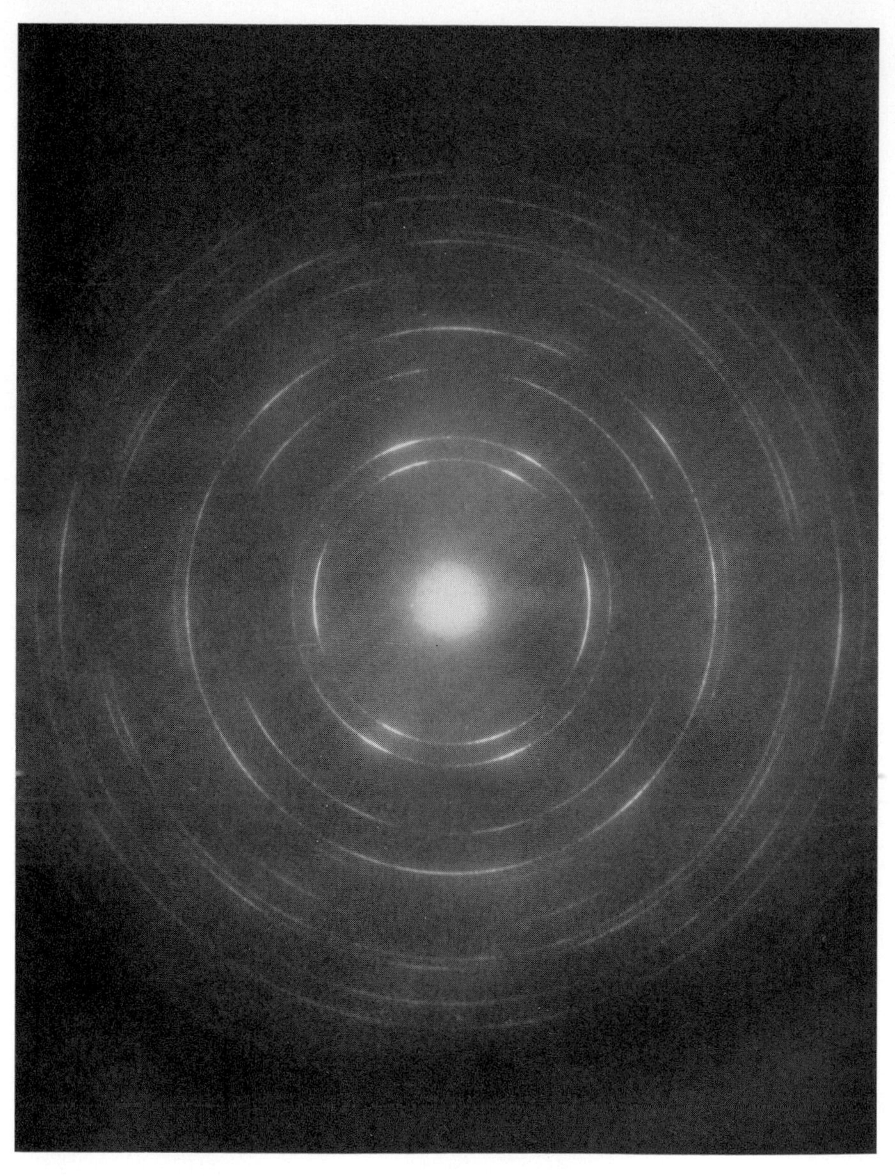

Plate 3. Inclined thin film texture pattern showing arced Debye rings.

Plate 4. Inclined thin film texture pattern showing ellipses.

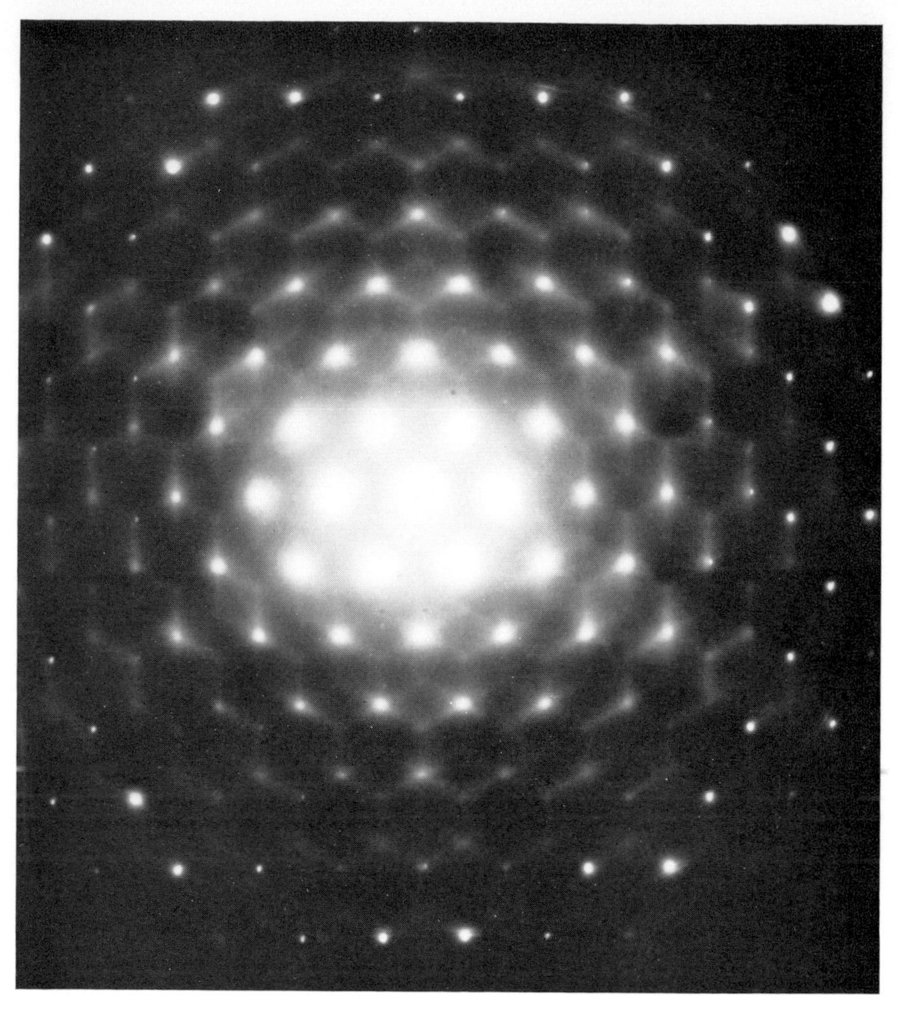

Plate 5. Streak transmission pattern of silicon with the electron beam directed along the 111 direction.

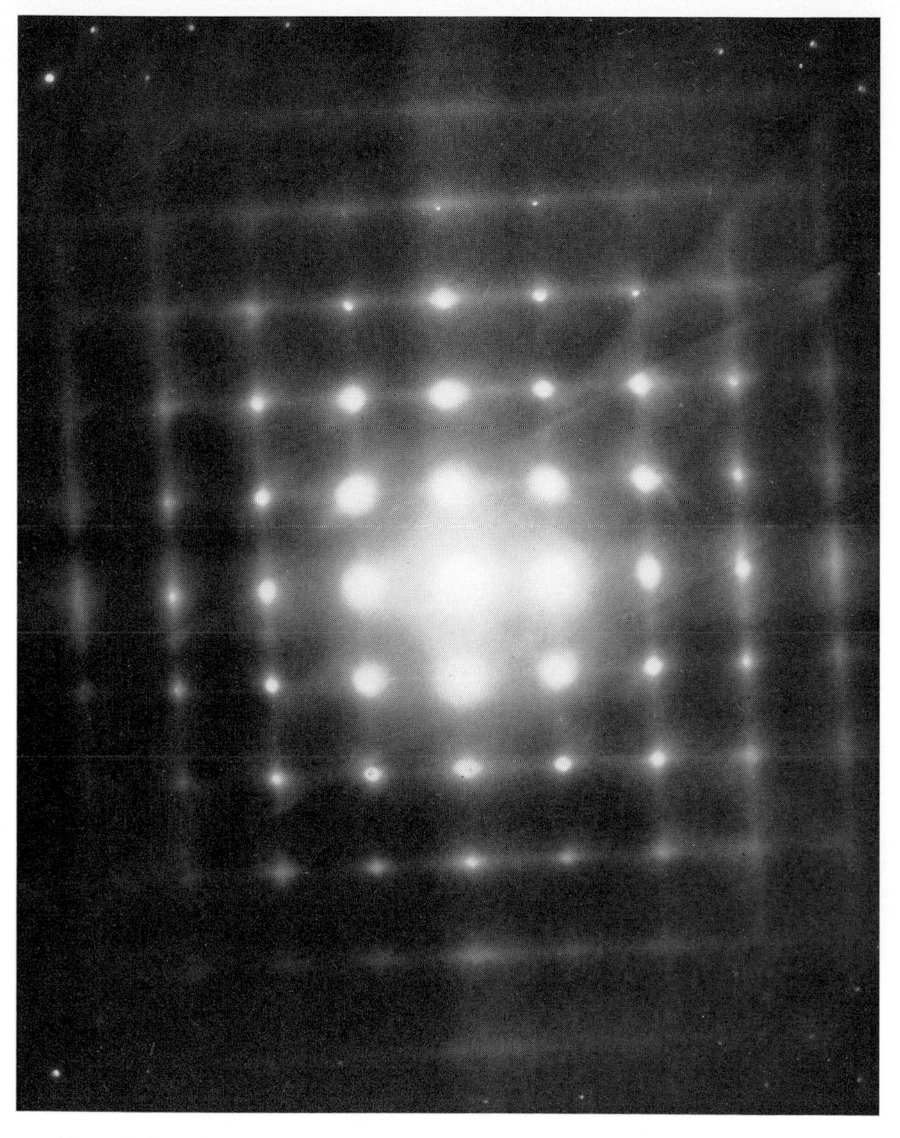

Plate 6. Streak transmission pattern of silicon with the electron beam directed along the 100 direction.

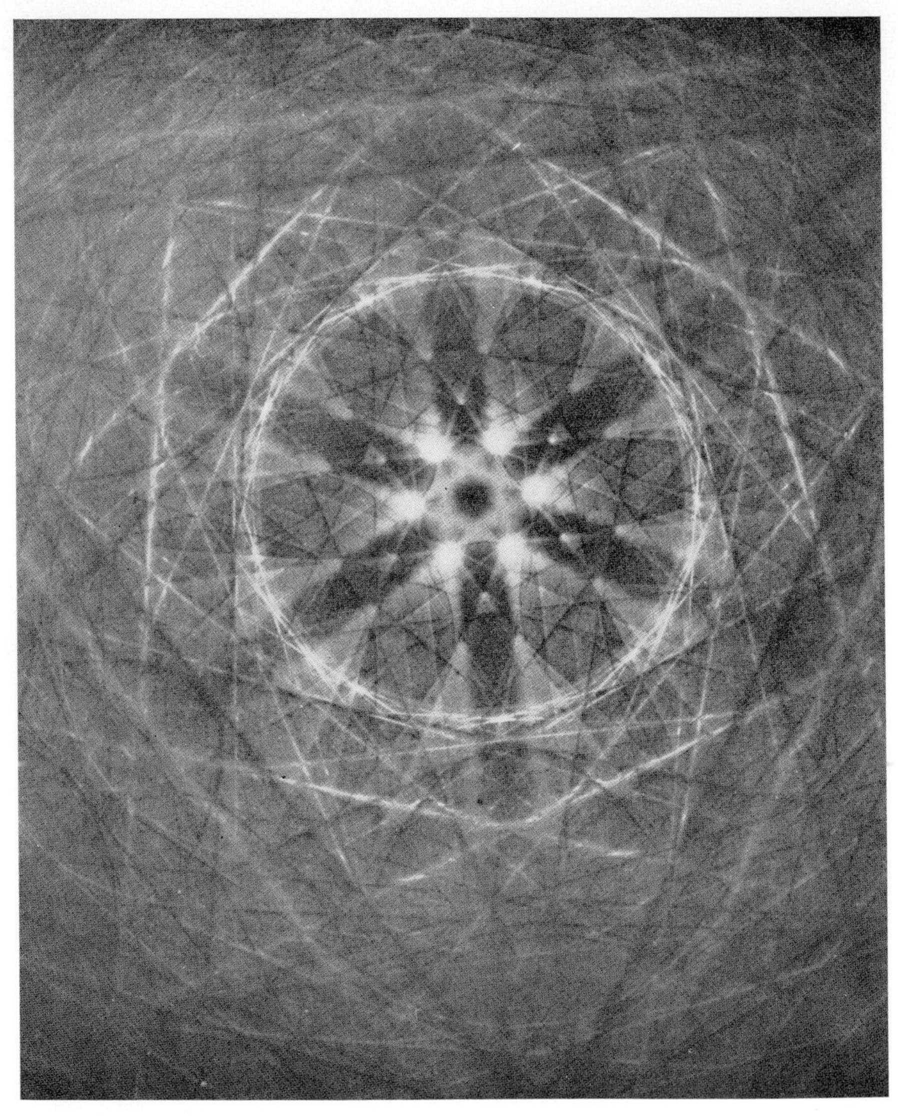

Plate 7. Transmission pattern of germanium showing Kikuchi lines and bands.

respect to one another so that a point of one plane lies vertically over the centre of a square outlined by four points of the adjacent plane.

Using the relations between the crystal lattice and the reciprocal lattice discussed in § 3.3 (cf. equation 3.16 and Fig. 3.7), we find that the crystal lattice must be a *face-centred cubic* structure with a unit cell of side 500 pm.

The somewhat idealized example studied above indicates how a single-crystal pattern can be interpreted. The positions of the spots in the pattern can be measured very accurately. Hence the mesh size of the points of the reciprocal lattice lying in a plane perpendicular to the electron beam can, in favourable cases, be determined to better than 1 part in 10^4 (Kiendl [37]). The radii of the Laue zones are much less well defined, and therefore the spacing, parallel to the electron beam, of the planes of the reciprocal lattice can hardly be measured to better than 1%.

4.3 *Determination of crystal size*

The size of a crystal cannot readily be determined from an electron diffraction pattern, and whenever possible, other methods – e.g. electron microscopic observation of the crystal boundaries – are to be preferred. Where no other methods are available, the electron diffraction pattern can be used to give an estimate of crystal size to within about 30%.

We have seen in § 4.2, equation 4.2, that the width of the Laue zones of spots in a diffraction pattern gives a measure of the thickness of a single crystal in the direction of the beam. The accuracy of this estimation is limited by two facts. First, the extent of the zone is not sharply defined, the intensities of the spots diminishing as the zone boundary is approached. Secondly, if the crystal is slightly distorted, regions irradiated by the different parts of the primary beam will be oriented slightly differently. According to § 3.3, p. 54, this will result in the reciprocal lattice points being smeared out into small portions of spheres centred on the origin. The extent of such spherical caps may well be comparable with any elongation of the intensity regions of the reciprocal lattice into spikes due to the finite thickness of the crystal.

Estimates of crystal size are most valuable in the case of polycrystalline specimens, particularly when the crystallites overlap one another, so that their boundaries cannot be discerned in an electron micrograph. The mean size of the crystallites is given by equation 3.22a

$$N_1 = R/\Delta R$$

where N_1 is the average number of diffracting planes in a crystallite. R is the radius of the diffraction ring and ΔR is the integral width of the ring as defined in equation 3.21. If d is the spacing of the diffracting planes, the thickness t of the crystals is

$$\begin{aligned} t &= N_1 d = Rd/\Delta R \\ &= \lambda L/\Delta R \end{aligned} \qquad 4.5$$

from 1.9.

In measuring ΔR, allowance has to be made for the finite width of the primary beam at the photographic plate. It has been mentioned (cf. Fig. 3.9) that the intensity distribution across the ring is roughly Gaussian. If we have a primary beam of negligible width and carrying unit current, then the current density j in the diffracted beam at a distance r from the centre of the diffraction ring profile may be written

$$j = j_0 \exp \{-r^2/a^2\}. \qquad 4.6$$

The integral width of the diffracted beam is

$$\begin{aligned} w_0 &= \frac{1}{j_0} \int_{-\infty}^{+\infty} j \, dr. \qquad 4.7 \\ &= a\sqrt{\pi} \end{aligned}$$

It is found that the current distribution across a beam from an electron gun is roughly Gaussian. Hence if I_0 is the total current in the primary beam, the amount of current $I\,dr'$ in the portion of the beam distant r' to $(r' + dr')$ from the centre of the beam is given by

$$I = \frac{I_0}{b\sqrt{\pi}} \exp \{-r'^2/b^2\}. \qquad 4.8$$

The integral width of the primary beam is then

$$w_0 = b\sqrt{\pi}. \qquad 4.9$$

The portion $I\,dr'$ of the primary beam will produce in the diffracted beam an intensity distribution

$$Ij\,dr' = \frac{I_0 j_0\,dr'}{b\sqrt{\pi}} \exp -\{r'^2/b^2 + (r - r')^2/a^2\}.$$

The whole primary beam will therefore produce a distribution of intensity in the diffracted beam

$$\int_{-\infty}^{+\infty} Ij\, dr' = \frac{I_0 j_0}{b\sqrt{\pi}} \exp -\{r^2/(a^2+b^2)\}$$

$$\times \int_{-\infty}^{+\infty} \exp -\left\{\frac{a^2+b^2}{a^2b^2}\left(r' - \frac{rb^2}{a^2+b^2}\right)^2\right\} dr'$$

$$= \frac{I_0 j_0 a}{(a^2+b^2)^{\frac{1}{2}}} \exp -\{r^2/(a^2+b^2)\}. \qquad 4.10$$

The diffracted beam therefore has an integral width

$$w' = [\pi(a^2+b^2)]^{\frac{1}{2}}. \qquad 4.11$$

Hence from 4.5, 4.7, 4.9 and 4.11, the thickness of the crystals is

$$t = \frac{\lambda L}{(w'^2 - w_0{}^2)^{\frac{1}{2}}}. \qquad 4.12$$

In practice, integral widths are not particularly easy to measure. It is, however, fairly easy to determine from a microdensitometer trace of the diffraction pattern the width of a diffraction ring or central beam at which the intensity is half the maximum (Fig. 4.3). For a Gaussian intensity distribution

$$i \propto \exp\{-r^2/\sigma^2\}$$

this 'half value width' W is

$$W = 2\sigma(\ln 2)^{\frac{1}{2}}$$

whereas the integral width of the same distribution is

$$w = \sigma\sqrt{\pi}.$$

Hence, in terms of half widths, equation 4.12 may be written

$$t = 2\left[\frac{\ln 2}{\pi}\right]^{\frac{1}{2}} \frac{\lambda L}{(W'^2 - W_0{}^2)^{\frac{1}{2}}}$$

$$= 0{\cdot}9394 \frac{\lambda L}{(W'^2 - W_0{}^2)^{\frac{1}{2}}} \qquad 4.13$$

The quantity t calculated from 4.13 is (cf. p. 62) the average of the lengths of columns, perpendicular to the reflecting plane, into which the crystals may be imagined to be divided; each column being weighted proportionally to its volume. A convenient measure of the size of a crystal is a length D equal to the cube root of the volume

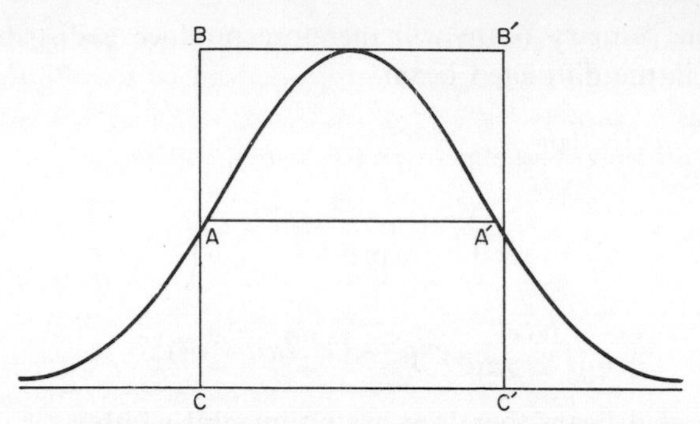

Fig. 4.3. AA′ is the half value width of the intensity distribution. For a Gaussian distribution, this is 6% less than the integral width BB′ which makes the area of the rectangle BB′C′C equal to the area under the curve.

of the crystal. The relation between D and t has been given by Stokes and Wilson [62] for crystals in the form of regular solids. In general $D = Kt$ where K is a constant lying between 1 and 1·2. Combining this with 4.13 we see that, to the accuracy of this theory

$$D = \lambda L/(W'^2 - W_0^2)^{\frac{1}{2}}. \qquad 4.14$$

There are three important sources of error in the determination of the average size D of the crystals according to 4.14. The first of these, the uncertainty in the value of K, has already been mentioned. The second is that the relation

$$W'^2 = W^2 + W_0^2, \qquad 4.15$$

which gives the correction to be applied to the measured width of a diffraction ring to allow for the finite width of the primary beam, may not be valid if the intensity distributions are not Gaussian. For example, if they are represented by curves of the form

$$I \propto (1 + r'^2/a^2)^{-1}$$

it can be shown that 4.15 should be replaced by

$$W' = W + W_0 \qquad 4.15a$$

with a corresponding change in 4.14. It is extremely difficult to determine experimentally the exact form of the intensity distribution, and hence to be sure whether 4.15 or 4.15a is valid. The difference between these two expressions can lead to serious error when W is small, i.e. when the crystals are large.

The third source of error is that elastic strains in the specimen can broaden the diffraction rings, and hence lead to an under-estimate of the size of the crystals. In consequence, it is unwise to rely on crystal size measurements to nearer than about 30%.

4.4 *Texture of polycrystalline specimens*

A very common type of specimen consists of a mass of crystallites oriented so that a particular crystallographic axis has roughly the same direction for every crystallite. For example, thin films prepared from metals having face-centred cubic crystals often have the [100] crystal direction normal to the plane of the specimen, but the orientation of the crystals is otherwise random. Again, metal sheet prepared by rolling often has the crystals oriented so that a certain crystal direction lies in the plane of the sheet and parallel to the direction of rolling. Such specimens are known as *textures*, and the common crystallographic direction of all the crystals is the orientation axis of the texture. Since the physical properties of a solid are often very sensitive to crystal orientation, the examination of textures is of considerable practical importance.

4.4.1 THIN FILM TEXTURES

To understand the general appearance of a diffraction pattern from a thin film texture, let us consider the case of a face-centred cubic material having one of the cube axes normal to the plane of the specimen. Figure 4.4 shows part of the reciprocal lattice of a crystal of such a specimen. The nature of the texture requires that the [001] direction be vertical, but there is no other restriction on the orientation of the reciprocal lattice. The superposition of the patterns from many crystals is equivalent to a rotation of the reciprocal lattice about the [001] direction, and results in the *spots* being replaced by *circles* as shown in the figure. If the specimen is normal to the electron beam, the diffraction pattern will be the section of this figure in reciprocal space by the plane ABCD (§ 3.3), which is normal to the direction EF of the electron beam and the orientation axis of the texture. Obviously the pattern consists of concentric circles, like the pattern from a polycrystalline specimen with complete random orientation of the crystallites. We note, however, that some of the diffraction rings are missing; in this example the only diffraction rings are those with indices of the form $h\,k\,0$. If the specimen is tilted, the pattern becomes a section of the reciprocal space figure by a plane such as ABCD in Fig. 4.5. Instead of *rings* we have *pairs of spots* formed by the intersection of the plane ABCD with the circles.

If the orientation of the crystallites is not perfect, so that the [001] direction lies within a cone of semi-angle α about the normal to the film, then the circles of Fig. 4.5 are broadened into spherical zones of angular width 2α. The intersection of these with a plane gives *arcs of circles* instead of *spots*. Plate 3 is an example of such a pattern.

The characteristics of the diffraction pattern of a thin film texture

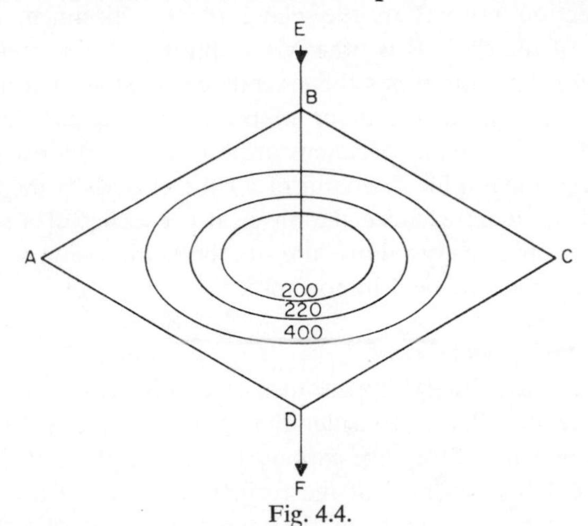

Fig. 4.4.

are therefore that (*a*) when the film is normal to the beam the pattern consists of rings, but the relative intensity of these rings is different from those produced by a polycrystalline specimen and (*b*) when the film is tilted the rings break up into arcs (spots, in the case of exact orientation).

The diffraction pattern from a tilted thin film texture with near perfect orientation therefore consists of spots in place of the circles which a specimen with random orientation would give. Frequently, these spots are seen to be arranged in regular curves. A very common pattern consists of spots arranged in *ellipses*. The origin of these ellipses is as follows. Thin film textures often have the most densely packed planes of atoms in the plane of the film. It therefore follows that the spacing between these planes in the direction of the normal to the film is a maximum. In reciprocal space, this corresponds to a small spacing between reciprocal lattice points in this direction. The reciprocal lattice of a single crystallite must therefore consist of rows of closely spaced points normal to the film, the distances between the rows being relatively large. The reciprocal lattice of the texture,

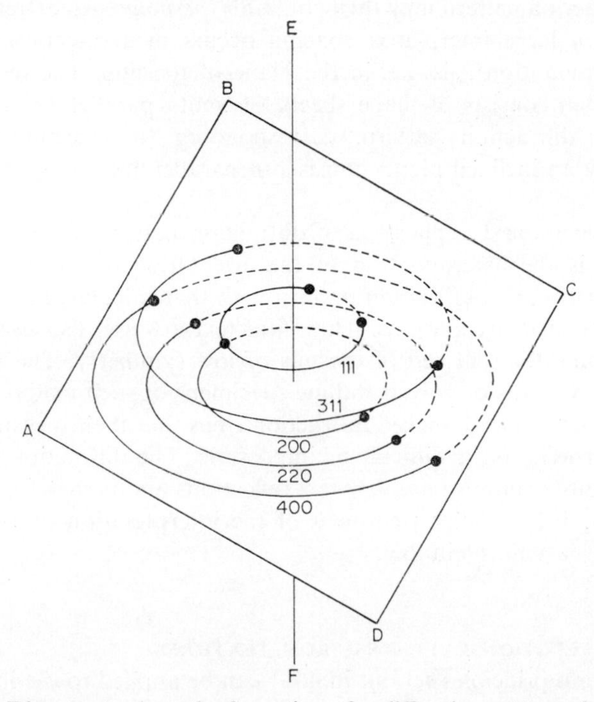

Fig. 4.5. Diagram to show the formation of a diffraction pattern of a texture inclined to the electron beam. EF is the orientation axis of the texture, so that the reciprocal lattice consists of circles centred on this line. The electron beam is normal to the plane ABCD.

formed by rotating this about an axis normal to the film, therefore consists of circles lying close together on cylinders, the separation between the various cylindrical sheets being fairly large. The section of this figure by an inclined plane will obviously produce a set of ellipses. If the film is very thin, then the reciprocal lattice points are elongated into spikes normal to the film (§ 3.4) and the reciprocal lattice of the texture consists of continuous cylindrical sheets. The corresponding diffraction pattern then consists of almost continuous ellipses (Plate 4).

Occasionally textures have a particularly large interplanar spacing in a direction inclined to the orientation axis. An obvious modification of the argument of the previous paragraph shows that the reciprocal lattice will then consist of *cones*. The plane section corresponding to the diffraction pattern may then contain *hyperbolae*. An extreme case is when a large interplanar spacing occurs in a direction normal to the orientation axis, i.e. in the plane of the film. The reciprocal lattice then consists of plane sheets of spots parallel to the film, and the diffraction pattern, corresponding to a section of this figure by an inclined plane, consists of parallel lines of spots (layer lines).

The commonest application of diffraction patterns from thin film textures is the determination of the orientation of the crystallites by comparing the diffraction pattern with that calculated from various postulated orientations. Thin film textures are also useful for determining the unit cell of crystals of low symmetry. The Debye–Scherrer patterns of polycrystalline specimens of such materials contain so many closely spaced diffraction rings that their measurement and interpretation is difficult or impossible. The diffraction pattern of a texture contains relatively few reflections and is therefore easier to index. For a fuller treatment of the interpretation of thin film textures see Vainshtein [65].

4.4.2 REFLECTION PATTERNS FROM TEXTURES

The general principles set out in 4.4.1 can be applied to the interpretation of reflection patterns from textures. If the orientation axis of the texture is normal to the specimen surface, then the reflection pattern is similar to that from a thin film texture with its normal at a large angle to the beam; except, of course, that only half the pattern is visible.

If the orientation axis lies in the plane of the specimen, as is the case with rolled sheet metals, then the pattern will change if the specimen is rotated about an axis normal to its surface. When the electron beam is nearly parallel to the orientation axis, the pattern consists of continuous rings (or, rather, half rings) just like the pattern from a thin film texture set normal to the electron beam. If the specimen is rotated in azimuth about the normal to its surface, the continuous rings are broken up into spots or arcs like the pattern from an inclined thin film texture.

4.5 *Streak patterns*

Diffraction patterns from single-crystal specimens sometimes show, in addition to the usual net of spots, sets of linear streaks. There are three main types of pattern.

(*i*) Patterns with streaks due to refraction of the electron beam at sub-microscopic facets. This effect is more fully dealt with in Chapter 5.

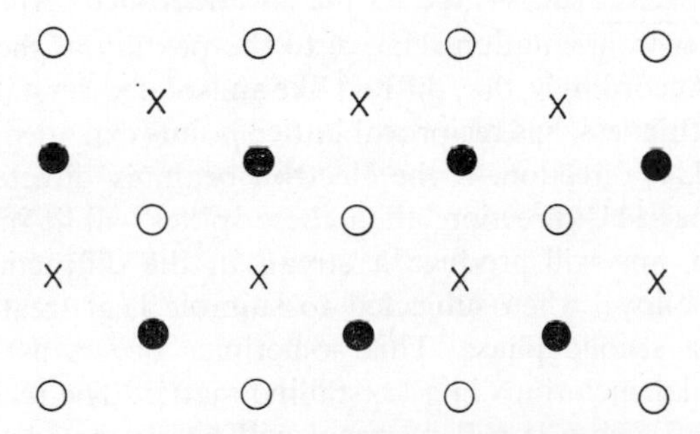

Fig. 4.6. 111 planes of face-centred cubic crystal. Successive planes of atoms are indicated by circles, crosses and dots.

(*ii*) Streaks due to reciprocal lattice points being extended into spikes owing to the finite size of the diffracting crystal (§ 3.4 and Plate 2). Two common types of structure can produce spikes in reciprocal space: (*a*) stacking faults, and (*b*) lamellar precipitates of a second phase in a matrix crystal.

(*a*) Stacking faults can occur in several types of crystal structure, of which the commonest is the face-centred cubic one found in many metals. If we examine a face-centred cubic lattice, we find that the (111) lattice planes have the arrangement of points shown in Fig. 4.6. Successive planes are arranged so that a point of one plane is over the centroid of a triangle formed by three points of the plane below. There are three types of plane, which differ in their lateral position. If we denote these by A, B, C, then the normal sequence of planes is

— ABC ABC —. It may happen in the growth of a crystal that this sequence is disturbed, giving an arrangement — ABCABCBACBAC —. This would produce a twin crystal: two crystals having a common (111) plane of atoms, but with one crystal rotated 180° with respect to the other about an axis normal to the twin plane. We can see this from an examination of Fig. 4.6. Suppose the plane of atoms represented by dots is the twin plane. The normal growth of the crystal would require the next plane of atoms be situated as indicated by the circles (A following C). A rotation through 180° about a vertical axis through the centre of the figure would convert this net of points into the net represented by crosses and would produce a stacking fault (B following C).

If we have two stacking faults in close succession: — ABCBACBACABC —, the six planes underlined form a thin slab which is in twin orientation relative to the portion of the crystal on either side. Accordingly, they diffract like an isolated crystal which, because of its thinness, has reciprocal lattice points extended into spikes along the [111] direction. If the electron beam is directed at right angles to the [111] direction, then these spikes will lie in the sphere of reflection, and will produce a streak in the diffraction pattern. (*b*) Certain alloys, when subjected to suitable heat treatment, may precipitate a second phase. This sometimes occurs as sub-microscopic lamellar inclusions in a crystalline matrix. The reciprocal lattice points of such a lamellar crystal will be elongated into spikes, and these will produce streaks in the diffraction pattern if the electron beam is correctly oriented, just as the twin crystal slab described in (*a*) above.

(*iii*) An essential characteristic of both kinds of streak pattern described in (*i*) and (*ii*) above is that they occur only for critical directions of the primary beam. Refraction streaks will occur only when the beam grazes a crystal facet. Streaks of the kind described in (*ii*) require the spikes in reciprocal space to be tangential to the sphere of reflection: in other orientations, the sphere of reflection will intersect the spikes in points, and the pattern will consist of spots displaced from the positions which they would occupy in the diffraction pattern of an infinite crystal. In contrast to these, the patterns to be described do not change greatly over a fairly large range of orientations of the primary beam relative to the crystal.

Plate 5 shows a typical pattern of this kind obtained from a thin silicon film set normal to the electron beam. The plane of the film is the (111) plane. We observe that the normal hexagonal array of spots is linked by a hexagonal network of streaks. This pattern persists with fair intensity if the specimen is tilted up to about 10°. If however we examine a film which is parallel to the (100) plane, we observe (Plate 6) a square array of spots linked by a square network of streaks. Again, it is found that this pattern persists if the specimen is tilted up to about 10°. Both these patterns can be interpreted in terms of *planes* in reciprocal space perpendicular to crystal directions of the type [110] (Honjo *et al.* [32]). A plane of intensity in reciprocal space is associated with a *line* of scattering points in the crystal in the direction of the normal to this plane.

The streaks in these diffraction patterns are therefore due to scattering by rows of atoms in the [110] type of direction. These are, in fact, the directions joining nearest-neighbour atoms. It is supposed that a prominent mode of the thermal vibrations of the crystal is one in which a whole row of atoms in the [110] direction is displaced as a unit along its length. Such displaced chains of atoms will scatter out of phase with the rest of the crystal, and therefore act as isolated scattering units. Confirmation of this theory is provided by the observation of Kitamura [38] that the intensity of the streak patterns diminishes when the temperature is lowered and the thermal vibrations are therefore reduced in amplitude.

The reason why this particular vibration mode is so prominent appears to be that it is produced by transverse acoustic vibrations polarized parallel to the atom chains (Komatsu [39]). Such vibrations have low frequency because, since the chains vibrate as a whole, there is little relative movement of nearest neighbours and therefore restoring forces are relatively small. Since, for a given energy of vibration, the amplitude is inversely proportional to the frequency, it is to be expected that this particular type of vibration will have a large amplitude. The large displacements of the atoms will produce a large diffuse scattering into the streaks in the pattern.

4.6 *Kikuchi and convergent beam patterns*

If a transmission pattern is obtained from a single crystal in the form of a fairly thick ($\sim 1 \mu$m) plate, then the usual spot diffraction pattern will be obscured by the intense, diffuse background produced by electrons which have suffered inelastic collisions. This diffuse background (Plate 7) is found to be crossed by a network of bright and dark lines, and is referred to as a Kikuchi pattern after its discoverer. Similar patterns can be observed in reflection if the electron beam is allowed to graze a plane surface of a large crystal.

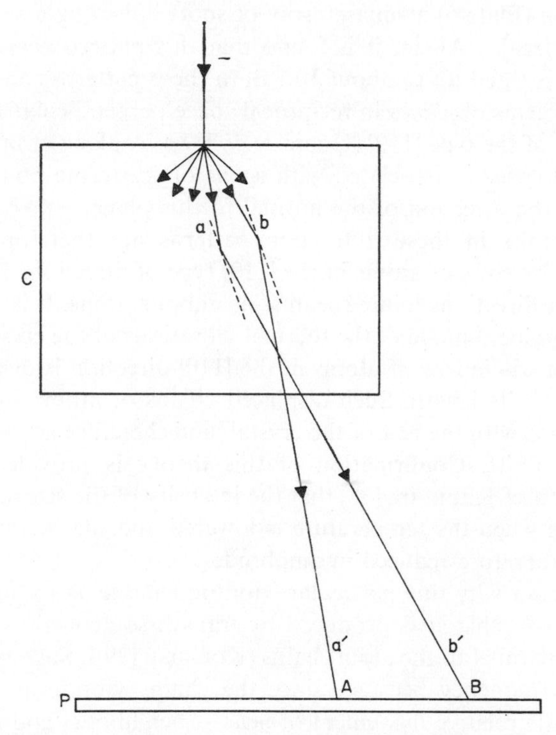

Fig. 4.7. Formation of Kikuchi pattern. The thickness of the crystal C is shown grossly exaggerated in relation to its distance from the photographic plate P.

The explanation of such patterns is as follows. A primary beam incident on the crystal is diffusely scattered* as indicated schematically in Fig. 4.7. Consider two rays *a* and *b* of this diffuse beam, which are inclined at the Bragg angle to a certain set of atomic planes which are indicated by the dotted lines. Then the beams *a* and *b* are

* The scattering is generally supposed to be inelastic, the electrons losing a few electron volts of energy, but the precise scattering mechanism is still uncertain.

diffracted and become the beams b' and a' respectively. The diffraction process therefore produces beams travelling in the same direction as the original ones (a' is parallel to a and b' to b). We may describe this by saying that the beams a and b are interchanged by diffraction. If these beams had the same intensity, this interchange would have no visible effect. In fact, the diffuse scattering is always strongest in the direction of the primary beam and diminishes in intensity as the scattering angle increases. Hence the beams a and b' are stronger than b and a'. Where the beam a' strikes the photographic plate at A we therefore have a rather weaker illumination than in the immediate vicinity, which is illuminated by diffuse radiation which has not undergone diffraction. Similarly, where the beam b' strikes the plate at B we have a stronger illumination than in the surrounding background.

Since the rays a' and b' are inclined at the Bragg angle θ to a reflecting plane, they are generators of cones – known as Kossel cones – of semi-angle $(\pi/2 - \theta)$, whose axes are the normal to the plane. The loci of the points A and B on the photographic plate is a plane section of these cones, i.e. an hyperbola. In fact, because the Bragg angle θ is small, the hyperbola appears as two parallel lines at A and B perpendicular to the plane of Fig. 4.7.

It is evident from Fig. 4.7 that if D is the distance between a pair of Kikuchi lines then, since θ is small

$$2d\theta = \lambda$$
$$D/L = 2\theta,$$

where, as usual, d is the interplanar spacing and L is the specimen-plate distance. Hence

$$Dd = \lambda L. \tag{4.16}$$

This theory explains the following observed features of the Kikuchi pattern:

(*a*) the pattern consists of pairs of dark and light parallel lines;
(*b*) the dark (defect) line is always closer to the point where the primary beam strikes the plate and the white (excess) line is always further away;
(*c*) the intersection of the reflecting plane with the photographic plate lies midway between the pair of lines;
(*d*) the distance apart of the pair of lines is given by equation 4.16.

Closely related to Kikuchi patterns are convergent beam (or Kossel–Möllenstedt) patterns. These are obtained by increasing the current through the focusing lens so that the incident beam is focused on a thin crystal as shown in Fig. 4.8. Since the crystal is traversed by a cone of rays, the same diffraction effects occur as described above, and result in a similar pattern on the photographic plate. In fact, the only difference between a Kikuchi and a convergent beam pattern is that in the former the cone of rays traversing the crystal is produced by diffuse scattering and in the latter by external electron optical means.

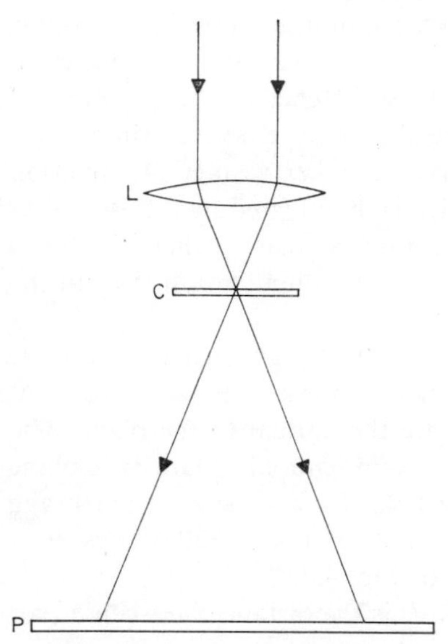

Fig. 4.8. Formation of convergent beam pattern. L Focusing lens; C Crystal; P Photographic plate.

The most important application of Kikuchi or convergent beam patterns is the determination of the precise orientation of a crystal. Obviously two pairs of lines, giving the intersection of two crystal planes with the photographic plate, completely determine the orientation of the crystal. The orientation of a crystal can be found much more precisely from a Kikuchi pattern than from an ordinary spot pattern because, as will be shown in Chapter 6, a more exact treatment of diffraction in terms of the dynamical theory shows that the diffracted beams are changed but little in direction or intensity if the orientation of the crystal with respect to the incident beam is altered by an amount comparable with the Bragg angle. The positions of Kikuchi lines, on the other hand, depend on the orientation of the crystal precisely as Fig. 4.7 indicates. Consequently, if both ordinary diffraction spots and Kikuchi lines exist in the same pattern, it is found that a small rotation of the crystal causes the latter to move over the stationary pattern of spots.

The simple theory given above fails to account for two important features of Kikuchi and convergent beam patterns. In the first place, if in Fig. 4.7 the beams *a* and *b* are symmetrically situated with respect to the primary beam – this will occur when the direction of the primary beam lies in a crystal plane – they will be equally

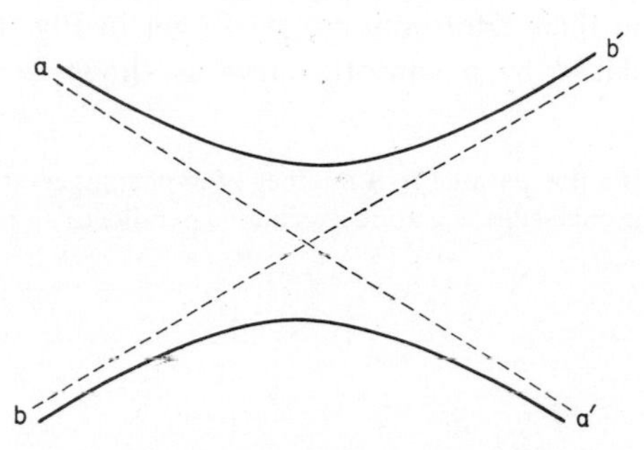

Fig. 4.9. Where the elementary kinematic theory of Chapter 3 predicts two intersecting Kikuchi lines aa′ and bb′ (shown as broken lines), the more exact dynamical theory of Chapter 6 predicts two curves ab′ and ba′, which are shown as full lines.

intense, and their interchange will produce no effect. Hence one cannot have a pair of Kikuchi lines situated symmetrically with respect to the primary beam. Plate 7 shows that what is in fact observed under these conditions is a *band* (which may be either darker or lighter than the surroundings) occupying the space between what would otherwise be a pair of Kikuchi lines. The second feature

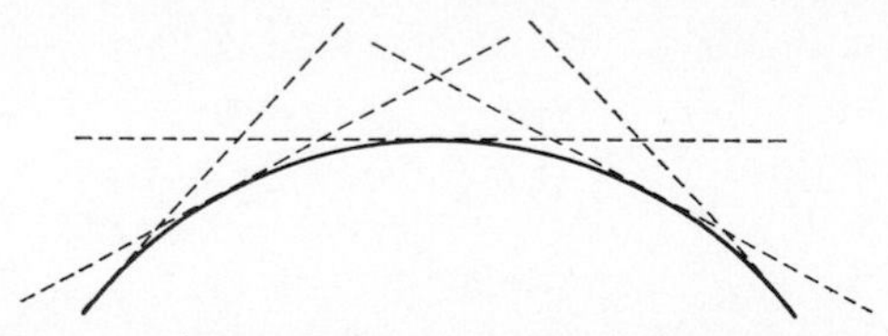

Fig. 4.10. Formation of Kikuchi envelope. Distortion of Kikuchi lines at their intersections results in a group of Kikuchi lines indicated by dotted lines being replaced by a smooth curve.

of the pattern is that when two Kikuchi lines intersect, they are distorted as indicated in Fig. 4.9. Both these effects are explicable in terms of the dynamical theory (6.51).

The distortion of Kikuchi lines at points of intersection appears to be responsible for the formation of *Kikuchi envelopes*. If the primary beam is nearly parallel to an important zone axis of the crystal,* then there will be a group of Kikuchi lines all tangential to a certain curve – their envelope. Because of the distortion of Kikuchi lines at their intersections, as shown in Fig. 4.9, the group of lines is replaced by a smooth curve as shown schematically in Fig. 4.10.

* A zone axis is a line parallel to a number of important crystal planes; e.g. in a cubic crystal the cube edge is a zone axis and is parallel to all planes of the type $(h\,k\,0)$.

5

Refraction Effects

5.1 *Refraction of electron waves*

A crystal consists of a regular array of atoms, each comprising a positive nucleus and a surrounding cloud of negative electrons. This regular, three-dimensional distribution of charge gives rise to an electrostatic potential within the crystal having the periodicity of the crystal lattice. We can therefore write

$$V = \sum_g V_g e^{2\pi i \mathbf{g} \cdot \mathbf{r}} \qquad 5.1$$

where the **g** are vectors of the reciprocal lattice. It is this periodic potential distribution which gives rise to diffraction, as will become

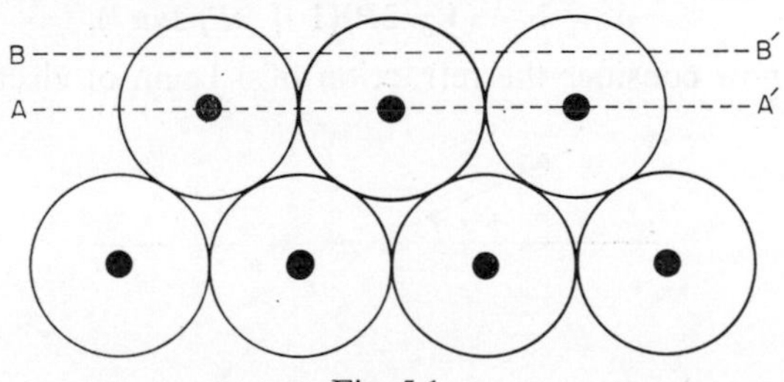

Fig. 5.1.

more immediately apparent in Chapter 6. However, there are not only periodic terms in the Fourier expression 5.1: there is also the zero-order, constant, term V_0 which corresponds to a mean potential within the crystal different from that of the surroundings. In practice, V_0 is always positive and of the order of 10 V. It is known as the *inner potential* of the crystal. The explanation, in general terms, of V_0 can be seen from Fig. 5.1, which represents the two layers of atoms next to the surface of the crystal. The black dots represent the positive nuclei and the shaded circular patches the negative clouds of electrons round each nucleus. It is obvious that there is an excess of positive charge in the plane AA′ and an excess of negative charge in a plane BB′ just outside it. These two planes are analogous to the electrodes of a charged parallel plate condenser. There will be an electrostatic field directed from AA′ to BB′. Hence the potential at AA′, and points lower than this within the crystal, will be greater than at points above BB′, outside the crystal.

We see at once from equation 1.3 that there will be a change of wavelength when an electron beam enters or leaves a crystal: the crystal will therefore refract an electron beam just as an optically dense medium refracts a beam of light. If P is the accelerating potential of the electrons, their kinetic energy will be eP outside the crystal and $e(P + V_0)$ inside. Hence, from 1.3 the refractive index of the crystal will be

$$\mu = \lambda_{\text{vacuum}}/\lambda_{\text{crystal}} \qquad 5.2$$

$$= [(P + V_0)/P]^{\frac{1}{2}}$$

$$\approx 1 + V_0/2P \qquad 5.3$$

since $V_0/P \sim 10^{-3}$ or less for fast electrons. If the more exact relativistic expression 1.4 is used, we have, to the same approximation,

$$\mu = 1 + (V_0/2P)(1 + eP/2mc^2). \qquad 5.4$$

Let us now consider the refraction of a beam of electrons at the

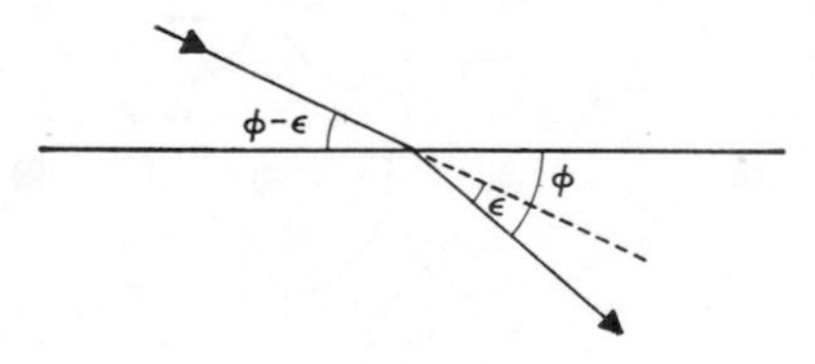

Fig. 5.2.

surface of the crystal. If ϕ is the grazing angle of the beam inside the crystal (Fig. 5.2), then Snell's law of refraction states that

$$\cos(\phi - \varepsilon) = \mu \cos \phi.$$

Since 5.3 shows that μ is always nearly unity, and therefore that ε is small, we may write

$$\varepsilon = (\mu - 1) \cot \phi$$

or

$$\varepsilon = \frac{V_0}{2P} \cot \phi \qquad 5.5$$

from 5.3.

Snell's law can be written in another form. If $k_0 = 2\pi/\lambda_0$ and $k = 2\pi/\lambda$ are the magnitudes of the wave vectors in vacuum and in the crystal, λ_0 and λ being the corresponding wavelengths, then

$$\mu = k/k_0$$

and
$$k_0 \cos(\phi - \varepsilon) = k \cos \phi.$$

Hence *the component of the wave vector tangential to the surface of the crystal is unchanged when the wave enters the crystal.* This comes about because it is necessary that there be no discontinuity in the wave function ψ at the crystal boundary.

If L is the distance of the photographic plate from the specimen, then the deviation of the point where the beam strikes the plate which is produced by refraction at a single surface is

$$\Delta_1 = L\varepsilon = \frac{LV_0}{2P} \cot \phi.$$

Since the refracted ray lies in the plane of incidence, the direction of the deflection is that of the projection on to the photographic plate of the normal to the crystal surface.

If the beam passes through a *wedge-shaped* crystal the resultant deviation is the vector sum of deviations Δ_1 and Δ_2 produced by the entrance and exit faces.

The preceding calculation is strictly appropriate only for the main, undiffracted beam. The diffracted beams will have grazing angles with the exit face, and consequently values of Δ_2, slightly different from that of the undiffracted beam. However, this difference is only $\sim$1% for electrons of about 50 keV energy,* and produces a slight increase in the size of the diffraction pattern.

EXAMPLE

Show that, if the exit surface of the crystal is nearly normal to the electron beam, the increase in the size of the diffraction pattern due to refraction exactly compensates the reduction in size due to the decrease in the wavelength of the electrons produced by the inner potential of the crystal.

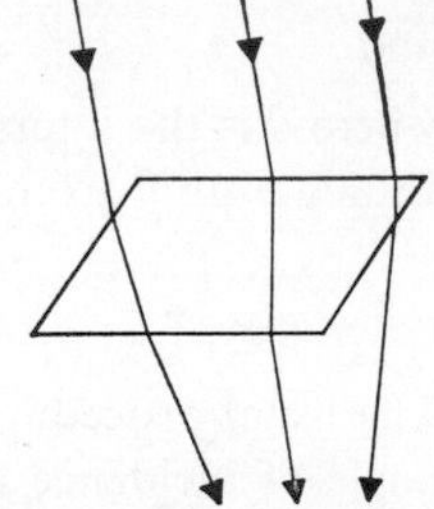

Fig. 5.3. Refraction by a polyhedral crystal.

In general, we have to deal with polyhedral crystals, and therefore the main beam is refracted as indicated schematically in Fig. 5.3. The central spot of the pattern, and each diffracted spot, is replaced

*Except for the special case dealt with in § 5.2.

by a similar group of spots produced by refraction at the various crystal facets. Using 50-keV electrons and a camera length of 500 mm, the separation of these spots is only of the order of 50 μm. It can therefore be observed only in a high resolution camera, and even then it is difficult to measure accurately. However, Cowley and Rees [15] were able to obtain moderately accurate values of the inner potential of MgO and CdO from an analysis of the multiple spots in the diffraction patterns of smokes of these materials.

5.2 *Measurement of inner potential*

Equation 5.5 shows that the deviation produced by refraction becomes larger when the electron makes a small glancing angle with the surface. Hence most measurements of inner potential have been made by studying reflection patterns from atomically smooth crystal faces. Consider the beam which enters and leaves the same face of a

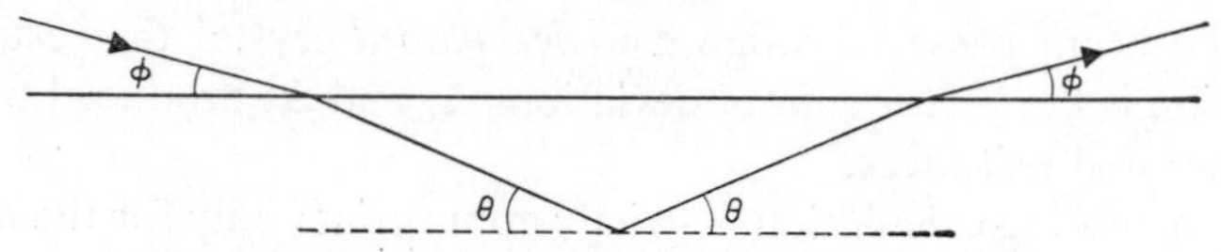

Fig. 5.4. Effect of refraction on a 'reflection' pattern.

crystal at glancing angle ϕ and is reflected at a Bragg angle θ by atomic planes parallel to the surface (Fig. 5.4). Then

$$\cos \phi = \mu \cos \theta$$

and

$$2d \sin \theta = n\lambda,$$

where d is the interplanar spacing and n the order of the diffraction. Eliminating θ from these equations, we obtain, since $\mu \approx 1$,

$$\sin^2 \phi = \frac{n^2\lambda^2}{4d^2} - (\mu^2 - 1). \qquad 5.6$$

The usual procedure is to rotate the crystal slowly so as to vary the angle of incidence over a few degrees of grazing incidence, so that the reflections of various orders appear in turn and produce a row of spots on the photographic plate. The values of ϕ are determined by measuring the distance between each diffracted spot and the spot due to the incident beam. If, then, $\sin^2 \phi$ is plotted against n^2, a straight line is obtained with intercept $-(\mu^2 - 1)$ on the $\sin^2 \phi$ axis. The inner potential V_0 can then be calculated from 5.3.

6

Dynamical Theory of Electron Diffraction

6.1 *Introduction*

In Chapter 3, we calculated the intensity of the electron scattering by a crystal. It was assumed that the wave incident on each atom is simply the primary wave falling on the crystal, and therefore that the total amplitude of the diffracted wave is proportional to the number of atoms in the crystal. This is not quite correct, for it is clear that the amplitude of the original wave will be diminished as it passes successive reflecting planes of atoms, owing to the loss of electrons into the reflected wave. This attenuation of the original wave (known as extinction) is offset by the fact that the reflected beam is itself reflected by the same planes of atoms into the direction of the original beam. After passing through a sufficient thickness of crystal a *dynamical equilibrium* is established between the intensities of the original and the diffracted beams; the ratio of the amplitudes of these beams emerging from a thick crystal is independent of its thickness.

The existence of dynamical interaction effects in X-ray diffraction was recognized at an early stage in the development of this subject. Because atoms scatter electrons much more powerfully than X-rays, dynamical interaction effects are more important in electron than in X-ray diffraction. In X-ray diffraction, the intensity of the diffracted beam is accurately given by the kinematic theory and is proportional to the volume of the crystal provided this is less than about 0·1 mm thick. In electron diffraction, on the other hand, there is complete dynamical equilibrium between the original and the diffracted beam after passing through less than 10 nm of crystal; it is to be expected that for thicker crystals, the intensity of the diffracted beam will be independent of the size of the crystal. In this chapter, we shall trace some of the consequences of dynamical interaction between beams of electrons in a crystal.

There are several ways of tackling this problem. The most generally useful approach, which will be followed in this book, consists in setting up a solution of Schrödinger's equation in a form representing a primary and a diffracted wave in a crystal.

6.2 *The dynamical theory of refraction by a wedge-shaped crystal*

The electrostatic potential V in a crystal must necessarily have the periodicity of the crystal lattice. It can therefore be represented as a three-dimensional Fourier series

$$V = V_0 + \sum V_\mathbf{g}\, e^{2\pi i \mathbf{g}\cdot\mathbf{r}} \qquad 6.1$$

where g is a vector of the reciprocal lattice. Since V is a real quantity, the amplitudes $V_\mathbf{g}$, $V_{-\mathbf{g}}$ must be conjugate complex numbers, and can be written as $|V_\mathbf{g}|\, e^{i\delta}$ and $|V_\mathbf{g}|\, e^{-i\delta}$ respectively. V_0 is the inner potential of the crystal: the refraction effects associated with this have been discussed in Chapter 5.

Within the crystal, Schrödinger's equation becomes

$$\nabla^2\psi + \frac{8\pi^2 me}{h^2}(P' + \sum V_\mathbf{g}\, e^{2\pi i \mathbf{g}\cdot\mathbf{r}})\psi = 0, \qquad 6.2$$

where $P' = P + V_0$ and P is the accelerating potential applied to the electrons. It is convenient to separate the simple refraction effects due to V_0 from the more complex dynamical effects associated with the last term of 6.2. We may do this by supposing the crystal to be enveloped in a thin film of a continuum of inner potential V_0. On crossing a boundary between this film and vacuum, refraction occurs as described in § 5.1. Within the film and crystal, the electrons behave as if the accelerating potential were P', the electrostatic potential in the crystal being given by 6.1 with the term V_0 omitted.

The waves immediately outside the crystal (i.e. within the hypothetical thin film continuum) and inside the crystal can therefore be written as

$$\nabla^2\psi + k_0{}^2\psi = 0 \qquad 6.3$$

and

$$\nabla^2\psi + (k_0{}^2 + \sum_\mathbf{g} v_\mathbf{g}\, e^{2\pi i \mathbf{g}\cdot\mathbf{r}})\psi = 0 \qquad 6.4$$

respectively, where

$$k_0{}^2 = 8\pi^2 meP'/h^2 \qquad 6.5$$

and

$$v_\mathbf{g} = 8\pi^2 meV_\mathbf{g}/h^2. \qquad 6.6$$

A solution of 6.3 representing the incident wave is

$$\psi = e^{i\mathbf{k}_0\cdot\mathbf{r}},$$

where $\mathbf{k}_0$ is the wave vector of the incident wave and is of magnitude k_0. An analogous solution $\psi = e^{i\mathbf{k}\cdot\mathbf{r}}$, representing a single wave, will not satisfy 6.4 because the equation contains terms $e^{i(\mathbf{k}+2\pi\mathbf{g})\cdot\mathbf{r}}$ as well as $e^{i\mathbf{k}\cdot\mathbf{r}}$. This corresponds with the physical fact that in general a wave in the crystal is partially reflected by the crystal planes, and is therefore associated with other waves. We therefore try a solution of the form

$$\psi = \sum_{-\infty}^{+\infty} a_{\mathbf{g}}\, e^{i(\mathbf{k} + 2\pi\mathbf{g})\cdot\mathbf{r}} \qquad 6.7$$

representing a primary wave $a_0\, e^{i\mathbf{k}\cdot\mathbf{r}}$, which is the prolongation into the crystal of the vacuum wave $e^{i\mathbf{k}_0\cdot\mathbf{r}}$, and reflected waves of the form $a_{\mathbf{g}}e^{i(\mathbf{k}+2\pi\mathbf{g})\cdot\mathbf{r}}$. Substituting from 6.7 in 6.4

$$\sum_{\mathbf{g}}\sum_{\mathbf{h}} [a_{\mathbf{g}}\{k_0^2 - (\mathbf{k} + 2\pi\mathbf{g})^2\} + a_{\mathbf{h}}v_{\mathbf{g}-\mathbf{h}}]\, e^{i(\mathbf{k}+2\pi\mathbf{g})\cdot\mathbf{r}} = 0. \qquad 6.8$$

Equating to zero the coefficients of each harmonic term gives an infinite set of equations to determine the amplitudes $a_{\mathbf{g}}$. An exact solution of such a set of equations is impossible, but a sufficiently good approximation can be obtained. Let us first find the order of magnitude of the coefficients. We have

$$\frac{v_{\mathbf{g}}}{k_0^2} = \frac{V_{\mathbf{g}}}{P'}.$$

Now the Fourier coefficients of potential $\sim$ 10 V. Then if $P' \sim$ 50 kV

$$v_{\mathbf{g}}/k_0^2 \sim 2 \times 10^{-4}.$$

For an amplitude $a_{\mathbf{g}}$ to be appreciable, it is therefore necessary for the coefficient $k_0^2 - (\mathbf{k} + 2\pi\mathbf{g})^2$ in 6.8 to be of order $2 \times 10^{-4}\, k_0^2$.

The condition for reflection to occur according to the kinematic theory (chapter 3) is that $|\mathbf{k}| = |\mathbf{k} + 2\pi\mathbf{g}|$. It is the condition that the sphere of reflection shall pass through the reciprocal lattice point $\mathbf{g}$. The dynamical theory relaxes this requirement to some extent. Provided the difference $|\mathbf{k}| - |\mathbf{k} + 2\pi\mathbf{g}\,| \sim 10^{-4}\,|\mathbf{k}|$, $a_{\mathbf{g}}$ will still be appreciable and reflection will occur. It suffices if the distance of the reciprocal lattice point from the sphere of reflection is not more than about 10^{-3} of the radius of the latter.

In general, unless the crystal has a very special orientation with respect to the incident beam, not more than *one* lattice point will lie near the sphere of reflection. Hence only a_0 and one of the coefficients $a_{\mathbf{g}}$ will be appreciable; the others will be smaller by one or more orders of magnitude. If, in the set of equations formed by equating to zero the coefficients of the harmonic terms in 6.8, we delete all but the terms in a_0 and $a_{\mathbf{g}}$ we obtain

$$a_0[k_0^2 - k^2] + a_{\mathbf{g}}v_{-\mathbf{g}} = 0$$

$$a_0 v_{\mathbf{g}} + a_{\mathbf{g}}[k_0^2 - (\mathbf{k} + 2\pi\mathbf{g})^2] = 0. \qquad 6.9$$

Now
$$k_0^2 - k^2 = 2k_0(k_0 - k) - (k_0 - k)^2$$
$$\approx 2k_0(k_0 - k),$$

since $k_0 - k \sim 10^{-4}\, k_0$ for electrons with energy $\sim$50 keV. An analogous relation holds for $k_0^2 - (\mathbf{k} + 2\pi\mathbf{g})^2$. Hence 6.9 may be written

$$2k_0(k_0 - k) + \frac{a_{\mathbf{g}}}{a_0}v_{-\mathbf{g}} = 0$$

$$v_{\mathbf{g}} + 2k_0[k_0 - |\mathbf{k} + 2\pi\mathbf{g}|]\frac{a_{\mathbf{g}}}{a_0} = 0. \qquad 6.10$$

Now $v_{\mathbf{g}}$ and $v_{-\mathbf{g}}$ are conjugate complex quantities (since the electrostatic potential 6.1 is a real quantity). They can therefore be written as $ve^{i\delta}$ and $ve^{-i\delta}$ respectively. Hence from 6.10

$$\frac{a_{\mathbf{g}}}{a_0} = -A\, e^{i\delta}, \qquad 6.11$$

where
$$A = \sqrt{\frac{k_0 - k}{k_0 - |\mathbf{k} + 2\pi\mathbf{g}|}} \qquad 6.12$$

and
$$(k_0 - k)(k_0 - |\mathbf{k} + 2\pi\mathbf{g}|) = v^2/4k_0^2. \qquad 6.13$$

Equation 6.13 has the following geometrical interpretation. In Fig. 6.1, O is the origin of the reciprocal lattice and G is the lattice point concerned in the reflection. OG is the vector $2\pi\mathbf{g}$. a and b are portions of spheres of radius k_0 centred around O and G; for electrons of energy $\sim$50 keV, $k_0 \gg 2\pi g$ and these portions of spheres approximate to planes perpendicular to the plane of the figure. K defines the vectors KO, OG of the incident and diffracted waves according to the kinematical theory. On the dynamical theory, we also get a strong diffracted wave in a direction such as PG corresponding to a primary wave PO provided the perpendicular dis-

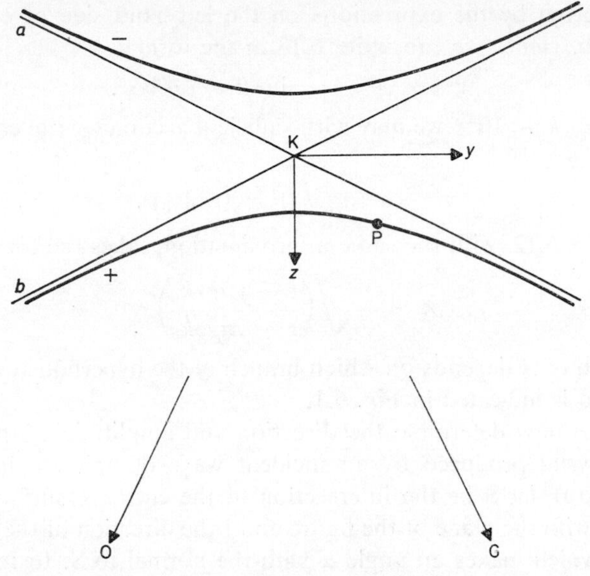

Fig. 6.1.

tances $k_0 - k$ and $k_0 - |\mathbf{k} + 2\pi\mathbf{g}|$ of P from a and b satisfy equation 6.13. The locus of P defined by this equation is known as the *dispersion surface* of the reflection **g**; it is approximately a hyperbolic cylinder having a and b as asymptotic planes. It is convenient to employ a rectangular system of coordinates with origin at K. The z axis bisects the acute angle OKG; since this angle is small ($\sim 10^{-2}$ radian), we may regard the z axis as nearly parallel to the electron beams. The y axis is parallel to OG, and the x axis is directed into the plane of the figure. Since the photographic plate will be placed normal to the beam, it is very nearly parallel to the xy plane and the y axis is parallel to the line from the centre of the diffraction pattern to the position of the diffracted spot according to the kinematic theory.

The equations of the asymptotic planes a and b are

a: $$z \cos \theta - y \sin \theta = 0 \qquad 6.14a$$

b: $$z \cos \theta + y \sin \theta = 0, \qquad 6.14b$$

where θ is the Bragg angle $\frac{1}{2}$OKG and $\sin \theta = \pi g/k_0$.

The perpendicular distances of a point xyz from the planes a and b are given by the expressions on the left-hand side of equations 6.14a, b. Hence we can write 6.13 in the form

$$z^2 \cos^2 \theta - y^2 \sin^2 \theta = v^2/4k_0^2.$$

Since $\pi g/k_0 \sim 10^{-2}$ we may with sufficient accuracy write $\cos \theta \approx 1$. Hence

$$z^2 - y^2(\pi g/k_0)^2 = (v/2k_0)^2. \tag{6.15}$$

Equation 6.12, with the same approximation, takes the form

$$A = \pm \sqrt{\left(\frac{z - y\pi g/k_0}{z + y\pi g/k_0}\right)}. \tag{6.16}$$

The sign of A depends on which branch of the hyperbolic cylinder P lies, and is indicated in Fig. 6.1.

Let us now determine the direction and amplitude of the waves in a crystal produced by an incident wave of unit amplitude. In Fig. 6.2(a), let S be the intersection of the entrance surface of the crystal with the plane of the figure and I the direction of the incident wave, which makes an angle α with the normal to S. In reciprocal space, Fig. 6.2(b) shows the incident ray K′O directed at a small angle ε = KOK′ to the direction required to satisfy the Bragg condition exactly. We now draw through K′ a line PP′ parallel to the normal to the entrance surface S. PP′ will in general be inclined to the plane of the figure. Then the points of intersection P, P′ of this line with the hyperbolic cylinder define the wave vectors PO, P′O of the primary waves in the crystal. For the fact that P lies on the hyperbolic cylinder means that the wave vector PO in the crystal satisfies the dynamical condition 6.13 for a strong reflection; and the fact that P lies on the normal to S through K′ means that the wave vector PO of the wave in the crystal, and the wave vector K′O of the wave outside the crystal, have the same component parallel to S, i.e. that the crystal wave PO and the external wave K′O match along S (*cf.* § 5.1). It is clear that P′O also defines a wave in the crystal.

The incident beam is split on entering the crystal into two primary beams of amplitude a_0, a_0' as indicated in Fig. 6.2(a). Some authors have expressed this by saying that the dynamical interaction of the primary and diffracted waves gives rise to double refraction. This conception is of limited utility, for the theory of double refraction in optics bears no relation to the phenomena studied here.

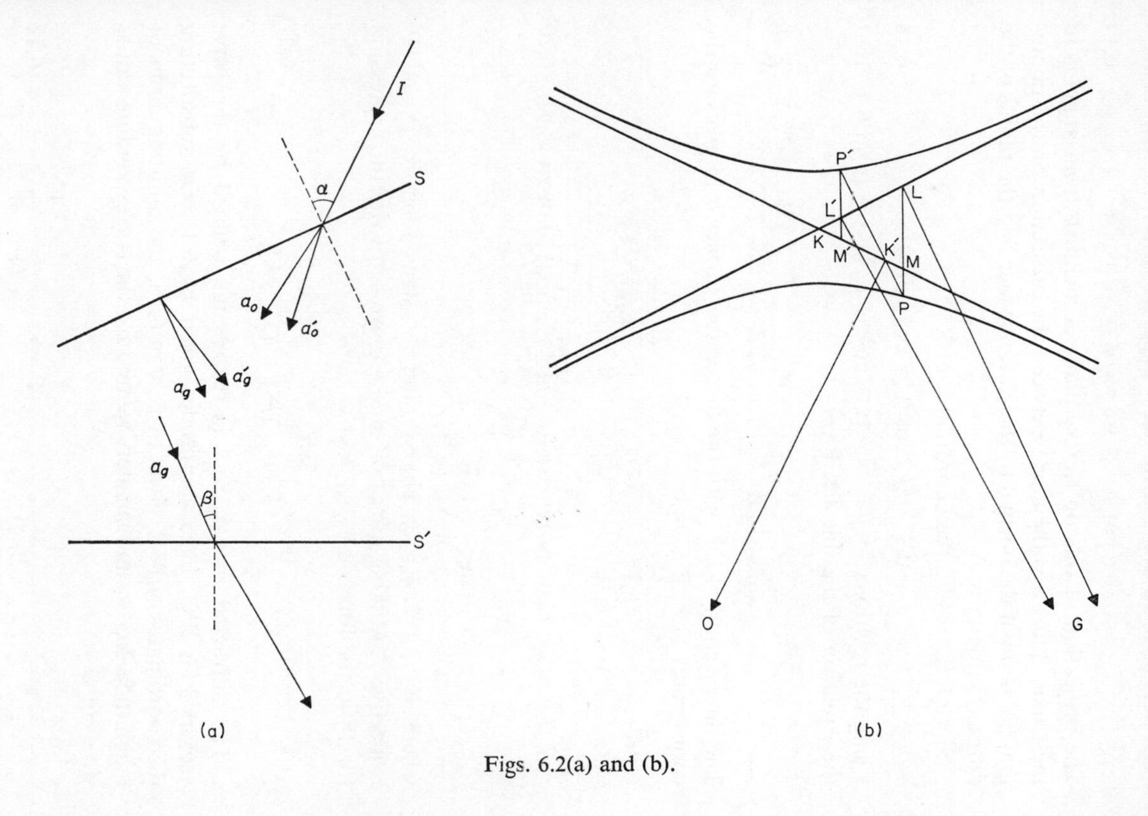

Figs. 6.2(a) and (b).

The line PP′ is inclined at angle α to the incident ray and therefore very nearly at an angle α to the z axis. Its direction cosines are

$$\sin\alpha\sin\phi \qquad \sin\alpha\cos\phi \qquad \cos\alpha. \tag{6.17}$$

The projection of this on the xy plane is inclined at an angle ϕ to the y axis. Thus ϕ is the angle between the projection on the photographic plate of the normal to the entrance face and the radius of the diffraction pattern.

Since KK′ $= \varepsilon k_0$, the coordinates of K′ are

$$0, \qquad \varepsilon k_0\cos\theta, \qquad \varepsilon k_0\sin\theta. \tag{6.18}$$

Using the relations $\sin\theta = \pi g/k_0$, $\cos \approx 1$, we see from 6.17 that the equation of the line PK′P′ is

$$\frac{x}{\sin\alpha\sin\phi} = \frac{y-\varepsilon k_0}{\sin\alpha\cos\phi} = \frac{z-\varepsilon\pi g}{\cos\alpha}. \tag{6.19}$$

The intersection of this with the dispersion surface 6.15 gives for the coordinates of P, P′

$$\begin{aligned} x &= \varepsilon\pi g\left[-1 \pm \sqrt{\left(1+\frac{1}{W^2}\right)}\right]\tan\alpha\sin\phi \\ y &= \varepsilon k_0 + \varepsilon\pi g\left[-1 \pm \sqrt{\left(1+\frac{1}{W^2}\right)}\right]\tan\alpha\cos\phi \\ z &= \pm\varepsilon\pi g\sqrt{\left(1+\frac{1}{W^2}\right)}, \end{aligned} \tag{6.20}$$

where $W = \varepsilon . 2\pi k_0 g/v$. The $+$ and $-$ signs refer to P and P′ respectively. Making use of 6.5 and 6.6, we can write the expression for W in the forms

$$W = \frac{\varepsilon\pi g}{k_0(V_{\mathrm{g}}/2P')} = \frac{\varepsilon\sin\theta}{(V_{\mathrm{g}}/2P')}. \tag{6.21}$$

The diffracted rays a_g, a_g' in the crystal, defined by the wave vectors PG, P′G, have amplitudes given by 6.11. On substituting the coordinates of P, P′ given by 6.20 in 6.16, and omitting terms of magnitude $2\pi g/k_0$ and smaller, we find for the corresponding values A, A' of A

$$A = \frac{1}{\sqrt{(1+W^2)}+W} \qquad A' = -\frac{1}{\sqrt{(1+W^2)}-W}. \tag{6.22}$$

Now in Fig. 6.2(a) the waves a_0, a_0' are the prolongation of the incident wave I of unit amplitude; there is no wave outside the *entrance* face of the crystal corresponding to the diffracted rays a_g, a_g'. The boundary conditions at the surface therefore require that

$$a_0 + a_0' = 1$$
$$a_g + a_g' = 0 \quad 6.23$$

From 6.11, 6.22 and 6.23 we obtain

$$a_0 = \tfrac{1}{2}\left[1 + \frac{W}{\sqrt{(1+W^2)}}\right] \qquad a_0' = \tfrac{1}{2}\left[1 - \frac{W}{\sqrt{(1+W^2)}}\right] \quad 6.24$$
$$a_g = -\frac{\tfrac{1}{2}e^{i\delta}}{\sqrt{(1+W^2)}} \qquad a_g' = +\frac{\tfrac{1}{2}e^{i\delta}}{\sqrt{(1+W^2)}}.$$

We therefore have two diffracted waves, each of intensity

$$1/4(1+W^2) \quad 6.25$$

travelling through the crystal in slightly different directions.

The waves a_0 and a_g in the crystal undergo refraction when they emerge from the surface S′ of Fig. 6.2(a). The direction of the emerging waves can be found by drawing a line PML, normal to S′, through the point P of Fig. 6.2(b) which defines the directions PO, PG of the crystal waves. If M and L are the intersections of this line with the spheres a and b of radius $k_0/2\pi$ centred on O and G, then MO and LG will be the directions of the emerging waves; for these wave vectors are each of magnitude $k_0/2\pi$ and have the same components normal to PML, i.e. in the exit surface S′, as the corresponding crystal wave vectors PO and PG. The same construction starting with the point P′ gives the emerging waves M′O and L′G corresponding to the crystal waves a_0' and a_g'.

Let the direction cosines of the normal to the exit surface S′ of the crystal be

$$\sin\beta\cos\phi' \qquad \sin\beta\sin\phi' \qquad \cos\beta \quad 6.26$$

defined in the same way as the corresponding quantities in 6.17 for the entrance face. Then making use of the expression 6.20 for the coordinates of the points P, P′, we can write down the equations of the lines PML and P′L′M′. Then using 6.14a, b we can find the coordinates of M, L, M′, L′. We can now compute the directions of the waves LG, L′G and find the amount by which the angle between the primary wave KO entering the crystal and either of these emerging diffracted waves deviates from the expectation of the simple kinematic theory. We shall not reproduce the tedious, though simple, algebra, but merely quote the final result. The diffracted spot undergoes a displacement due to the combined effect of simple refraction and dynamical refraction. If L is the specimen–plate distance, this displacement has a radial component

$$L(\tan\alpha\cos\phi - \tan\beta\cos\phi')\left(\frac{V_0}{2P} \pm \frac{V_g}{2P'}\sqrt{(1+W^2)}\right)$$

$$+ L(\tan\alpha\cos\phi + \tan\beta\cos\phi')\frac{V_g W}{2P'} \qquad 6.27$$

and a tangential component given by the same expression with $\cos\phi$, $\cos\phi'$ replaced by $\sin\phi$, $\sin\phi'$ respectively. Each of these diffracted waves has an intensity given by 6.25.

The dynamical theory of refraction summarized in 6.27 has been tested by Molière and Niehrs [48, 49], Altenhein and Molière [2] and Cowley, Goodman and Rees [14] using high resolution diffraction patterns of magnesium oxide smoke. This is a convenient substance for such studies because it crystallizes in well-formed rectangular parallelepipeds. This regular geometrical shape of the crystals simplifies the interpretation of the refraction patterns. We shall not discuss these studies further, merely remarking that they confirm the accuracy of equation 6.27, and lead to reasonable values for the Fourier coefficients of potential V_g.

6.3 *Dynamical diffraction in a crystal lamina*

The theory of § 6.2 can easily be adapted to the case of a crystal in the form of a parallel-sided slab. In Fig. 6.2(b), the normals to the entrance and exit surfaces now coincide, and the situation is as shown in Fig. 6.3. PK′LP′ is the common normal to the entrance and exit surfaces. The primary wave incident on the crystal, and the emerging undiffracted beam, are both represented by K′O. The corresponding waves in the crystal, of amplitude a_0 and a_0' respectively, are PO and P′O. The emerging diffracted wave is LG, and the corresponding crystal waves, of amplitudes a_g and a_g' respectively, are PG and P′G.

From 6.27, the diffracted spot is found to have radial and tangential displacements $(LV_gW/P')\tan\alpha\cos\phi$ and $(LV_gW/P')\tan\alpha\sin\phi$ respectively. More important is the fact that the two *pairs* of crystal waves a_0, a_0' and a_g, a_g' are each combined to form a *single* wave outside the crystal. Interference can therefore occur.

In Fig. 6.3, the wave vector K′O of the incident wave and the emerging undiffracted wave has a component $k_0 \sin\alpha$ perpendicular to PK′LP′, i.e. parallel to the surface of the crystal, and a component $k_0 \cos\alpha$ normal to the crystal (assuming $\varepsilon \ll \alpha$). The wave vectors PO, P′O of the corresponding crystal waves have the same tangential

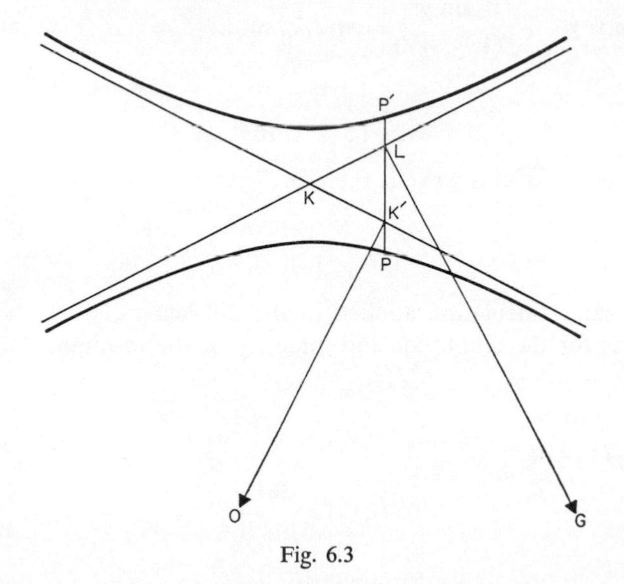

Fig. 6.3

components $k_0 \sin\alpha$, but their normal components are $k_0\cos\alpha - \mathrm{K'P}$ and $k_0\cos\alpha + \mathrm{K'P'}$ respectively. Using the expressions 6.18 and 6.20 for the coordinates of the points K′, P, P′, and remembering that $k_0 \sin\theta = \pi g$, we find for the magnitudes of K′P, K′P′ respectively

$$\varepsilon\pi g[\sqrt{(1 + 1/W^2)} \mp 1]\sec\alpha$$

Take a new coordinate system $\xi\eta\zeta$ with origin in the entrance surface and with the ζ axis directed normally into the crystal. Then, making use of the expressions 6.24 for the amplitudes a_0, a_0' of the crystal waves corresponding to the incident primary and emerging undiffracted wave, we can write these crystal waves in the form

$$\tfrac{1}{2}[1 + W/\sqrt{(1 + W^2)}] \exp i\{k_0 \sin \alpha\xi + (k_0 \cos \alpha + \varepsilon\pi g \sec \alpha)\zeta\} \exp\{- i\varepsilon\pi g\sqrt{(1 + 1/W^2)} \sec \alpha\zeta\}$$

and

$$\tfrac{1}{2}[1 - W/\sqrt{(1 + W^2)}] \exp i\{k_0 \sin \alpha\, \xi + (k_0 \cos \alpha + \varepsilon\pi g \sec \alpha)\zeta\} \exp\{+i\varepsilon\pi g\sqrt{(1 + 1/W^2)} \sec \alpha\zeta\}. \quad 6.28$$

These two waves combine to give a resultant wave of the form

$$T = \left[\cos \chi - \frac{iW \sin \chi}{\sqrt{(1 + W^2)}}\right] \exp i\{k_0 \sin \alpha\xi + (k_0 \cos \alpha + \varepsilon\pi g \sec \alpha)\, \zeta\}, \quad 6.29$$

where $$\chi = \varepsilon\pi g\sqrt{(1 + 1/W^2)} \sec \alpha\zeta. \quad 6.30$$

The intensity of this wave is therefore

$$|T^2| = \frac{W^2 + \cos^2 \chi}{1 + W^2}. \quad 6.31$$

The same calculation applied to the diffracted crystal waves a_g, a_g' gives for the amplitude and intensity of the resultant diffracted wave

$$G = \frac{i \sin \chi}{\sqrt{(1 + W^2)}} \exp i\{k_0 \sin \alpha'\xi + (k_0 \cos \alpha' + \varepsilon\pi g \sec \alpha)\zeta\} \quad 6.32$$

and $$|G|^2 = \frac{\sin^2 \chi}{1 + W^2}, \quad 6.33$$

where α' is the angle between the wave vector of the emerging diffracted wave and the normal to the crystal lamina. We note that the sum of the intensities of the undiffracted and diffracted beams is equal to the (unit) intensity of the incident primary beam.

The implication of equations 6.31 and 6.33 is that there is a continual interchange of electrons between the undiffracted and the diffracted beams. The periodic variation of the intensity of the two beams comes about because the two waves a_0, a_0' have slightly different wave vectors and therefore beat together; similarly the two waves a_g, a_g' form beats. The result summarized in equations 6.31 and 6.33 is often referred to as the *pendulum solution* of the wave equation.

The periodicity of the fluctuations in the intensities of the two beams is given by the change ($\delta\zeta$) in ζ which changes the argument of the sin/cos functions of 6.31 and 6.33 by π. This is given by

$$\delta = (\delta\zeta) \sec \alpha = \frac{W}{\varepsilon g\sqrt{(1 + W^2)}}. \qquad 6.34$$

δ is obviously the periodicity of the intensity variations measured along the general direction of the beams.

The value Δ of δ when $W = 0$, i.e. when the primary beam is directed so that the Bragg angle is exactly satisfied, is known as the *extinction distance*. From 6.21 and 6.34, we see that the extinction distance is

$$\Delta = \frac{W}{\varepsilon g} = \frac{\pi}{k_0(V_g/2P')}. \qquad 6.35$$

Using equation 1.3, the extinction distance can be expressed as

$$\Delta = \frac{h}{V_g}\sqrt{\left(\frac{P'}{2me}\right)} = \frac{\sqrt{(1{\cdot}5P')}}{V_g} \qquad 6.36$$

if V_g and P' are measured in volts and Δ in nm.

For electrons of about 50 keV energy, Δ is typically a few tens of nm. The concept of extinction distance is particularly useful in the interpretation of *extinction contours* in electron micrographs (Chapter 10).

If the crystal lamina has a thickness D, then the amplitudes and intensities of the transmitted and diffracted beams emerging from the far side of the lamina are given by equations 6.29–6.33 with $\zeta = D$. These are all functions of ε, the angle by which the primary beam deviates from the exact Bragg angle. Using 6.21, and writing

$$1/\delta^2 = S^2 = 1/\Delta^2 + (g\varepsilon)^2 \qquad 6.37$$

these equations can be written

$$G = \frac{i}{S\Delta}\sin(\pi SD\sec\alpha)\exp i\{k_0\sin\alpha'\xi + (k_0\cos\alpha' + \pi g\varepsilon\sec\alpha)D\} \qquad 6.38$$

$$|G|^2 = \frac{\sin^2(\pi SD\sec\alpha)}{S^2\Delta^2} \qquad 6.39$$

$$T = [\cos(\pi SD\sec\alpha) - i\sin(\pi SD\sec\alpha)(1 - 1/S^2\Delta^2)^{\frac{1}{2}}] \times \exp i\{k_0\sin\alpha\xi + (k_0\cos\alpha + \pi g\varepsilon\sec\alpha)D\} \qquad 6.40$$

$$|T|^2 = 1 - \frac{\sin^2(\pi SD\sec\alpha)}{S^2\Delta^2}. \qquad 6.41$$

Apart from a simple proportionality factor $(\pi \sec \alpha/\Delta b_1)^2$, equation 6.39 differs from equation 3.28b, which gives the intensity of the diffracted beam according to the kinematical theory, only in the replacement of εg by S. We saw in § 3.5 that equation 3.28b can be used to interpret convergent beam patterns from crystal laminae. Clearly, the dark fringes in the pattern correspond to values of S for which

$$SD \sec \alpha = m, \qquad 6.42$$

where m is an integer.

Writing $\varepsilon = x/L$ where x is the distance of a fringe from the centre of the diffracted patch and L is the specimen-to-plate distance

$$x^2 = \frac{m^2 d^2 L^2}{D^2 \sec^2 \alpha} - \frac{d^2 L^2}{\Delta^2} \qquad 6.43$$

Hence by plotting x^2 against m^2 we obtain a straight line from whose intercept on the x axis the extinction distance Δ and thus (using equation 6.36) Fourier coefficient of potential V_g can be determined.

The attractive possibility of measuring the Fourier coefficients V_g, from which the structure of an unknown crystal could be determined, by measuring *positions* of fringes on a photographic plate instead of *intensities* of rings or spots has, unfortunately, been realized in practice only to a very limited extent. The reason is that this method can be applied only to a very few crystals which can be obtained in the form of thin laminae a few hundred atom layers in thickness and a few mm in width.

6.4 *The intensity of the diffraction rings produced by a polycrystalline specimen*

Let us suppose we have a specimen consisting of a number of crystalline slabs in random orientation. If one of these crystals is to diffract electrons strongly, the normal to the reflecting planes of atoms must be inclined at a certain angle ϕ to the incident electron beam. If the Bragg condition is to be satisfied exactly, $\phi = \pi/2 - \theta$. According to the dynamical theory, the diffracted beam will have an appreciable intensity (given by 6.39) if the crystal is near to this orientation, so that $\phi = \pi/2 - \theta - \varepsilon$.

Now the probability of a crystal being oriented so that the normal to the reflecting planes makes an angle with the incident beam in the range ϕ to $\phi + d\phi$ is $1/4\pi$ times the solid between two cones of semi-angle ϕ and $\phi + d\phi$. This solid angle is $2\pi \sin \phi \, d\phi$, so that the probability is

$$\tfrac{1}{2} \sin \phi \, d\phi \approx \tfrac{1}{2} d\varepsilon$$

since θ and ε are both small if there is to be appreciable diffracted intensity. Combining this probability with the expression 6.39 for the intensity of the diffraction at this particular orientation, and using 6.35 and 6.37, we obtain for the ratio of the current i diffracted into the whole diffraction ring to the current i_0 in the incident beam

$$\frac{i}{i_0} = \frac{k_0 V_g}{4\pi g P'} \int_{-\infty}^{+\infty} \frac{\sin^2 A\sqrt{(1 + W^2)}}{1 + W^2} dW, \qquad 6.44$$

where $\quad A = k_0(V_g/2P')D \sec \alpha. \qquad$ 6.45a

Using 6.35, we can express the quantity A in terms of the extinction distance Δ

$$A = (\pi/\Delta)D \sec \alpha. \qquad 6.45b$$

The integral in 6.44 can be expressed in terms of the zero-order Bessel function

$$\int_{-\infty}^{+\infty} \frac{\sin^2 A\sqrt{(1 + W^2)}}{1 + W^2} dW = \pi \int_0^A J_0(2A) \, dA.$$

The integrated intensity of the ring is therefore

$$\frac{i}{i_0} = \frac{k_0 V}{4gP'} \int_0^A J_0(2A) \, dA. \qquad 6.46$$

If the crystals in the specimen are very thin, so that A is small, the integral in 6.46 becomes approximately equal to A. Then, using 6.45a

$$\frac{i}{i_0} = \frac{k_0^2 V_g^2}{8gP'^2} D \sec \alpha. \qquad 6.47$$

This expression implies that each crystal in the specimen contributes to the total diffracted current an amount proportional to $D \sec \alpha$, the distance traversed by the electrons in passing through the crystal. It follows that if we have a specimen consisting of several layers of crystals, we may replace the quantity $D \sec \alpha$ in 6.47 by t, the thickness of the specimen. Then, writing $1/g = d$, the interplanar spacing of the reflecting planes of atoms,

$$\frac{i}{i_0} = \frac{k_0^2 V_g^2 td}{8P'^2}. \qquad 6.48$$

When we come to compare this expression for the integrated intensity according to the dynamical theory with the 'kinematical' expression 3.18, we find that whereas the kinematical expressions are given naturally in terms of the structure factor E of the unit cell, the dynamical expression involves the Fourier coefficient of potential $V_{\mathbf{g}}$. The relation between these two quantities is given in Appendix 1. Making use of this relation, we can re-write 6.48 as

$$\frac{i}{i_0} = \frac{E^2\lambda^2 td}{2v^2} \qquad 6.49$$

which agrees exactly with the expression 3.18 derived on the kinematical theory.

For thicker specimens, the diffracted intensity falls below this 'kinematic' value because the integral in 6.46 is no longer approximately equal to A. In the limit, for very thick crystals, this integral becomes $\frac{1}{2}$ and in place of 6.48 we have, for a specimen consisting of a single layer of crystals,

$$\frac{i}{i_0} = \frac{E\lambda d}{4v}. \qquad 6.50a$$

If the specimen of thickness t consists of t/s layers of crystals each of average thickness s, then the total diffracted intensity will be t/s times the value given in 6.50a,

$$\frac{i}{i_0} = \frac{E\lambda d}{4v}\frac{t}{s}. \qquad 6.50b$$

The most important difference between the two expressions 6.49 and 6.50b for the total diffracted intensity on the kinematical and dynamical theories is that the former involves the *square* of the structure factor E and the latter the *first* power. If, therefore, we wish to determine the structure of an unknown crystal by comparing the intensities of the different diffraction rings so as to find the ratios of the structure factors, it is important to know whether the crystals in the specimen are thin enough for expression 6.49 to be a satisfactory approximation. Using 6.35 and A.1.6, we can express the ratio of the 'kinematical' expression 6.49 to the 'dynamical' expression 6.50b in the form

$$\frac{(i/i_0)_{\text{kinematic}}}{(i/i_0)_{\text{dynamic}}} = \frac{2\pi s}{\Delta}. \qquad 6.51$$

A possible criterion for the limit of validity of the kinematical theory would be that crystal thickness s which makes this ratio unity. Various other criteria have been suggested. They can all be summed up by stating that the crystal thickness must not exceed, in order of magnitude, the extinction distance Δ.

It might be supposed that the best way of determining the relative structure factors would be to produce specimens consisting of crystals large enough to ensure that the 'dynamical' expression 6.50b is valid. Unfortunately, there are some approximations in the dynamical theory developed here – notably the assumption that there is only *one* diffracted beam emerging from the crystal – so that 6.50b cannot be regarded as a satisfactory basis for the determination of structure factors.

We see that electron diffraction can best be used for structure determinations of crystals having a large extinction distance. If we combine equations 6.35 and A.1.6, we find for the extinction distance

$$\Delta = \pi v / E\lambda \qquad 6.52$$

Now E is a measure of the total amplitude scattered by all the atoms in the unit cell. If all the atoms scattered in phase, this amplitude would be proportional to the total number of atoms in the cell and, therefore, roughly proportional to the volume of the cell. The phase difference between waves scattered by atoms in different parts of the unit cell is larger the greater the dimensions of the cell. Hence E does not increase proportionally with v. We may therefore conclude that crystals with large unit cells will tend to have large extinction distances, and to be particularly suitable for structure determination by electron diffraction. Since both oxygen and hydrogen atoms have fairly low scattering factors, we may suppose that the presence of much water of crystallization will lead to low values of E and therefore large extinction distances. It is therefore not surprising that the most successful applications of electron diffraction have been in the field of complex hydrates. Finally, we note from 6.52 that the extinction distance is inversely proportional to the electron wavelength and therefore (1.3) directly proportional to the square root of the relativistic potential P^* used to accelerate the electrons. The use of high voltage diffraction apparatus for structure determinations is therefore obviously advantageous.

6.5 *Absorption effects*

In this section, we shall examine some of the effects of what is usually referred to as absorption. Strictly speaking, of course, an electron beam is not absorbed, for electrons can be destroyed only by positrons. 'Absorption' is the term given to inelastic scattering of electrons. Electrons which have suffered inelastic scattering have a different wavelength from those in the primary beam and, as a result of subsequent diffraction processes, are likely to be spread over the background of the final diffraction pattern; their contribution to the intensities of the diffraction rings is thereby lost. Similarly, in the electron microscope, inelastically scattered electrons are likely to be deflected so that they no longer pass through the objective lens diaphragm, and are therefore lost to the final image.

Absorption effects are best discussed in terms of what are known as Bloch waves. We have seen that a possible solution of the Schrödinger equations within a crystal is the *pair* of waves $a_0 e^{i\mathbf{k}\cdot\mathbf{r}}$ and $a_\mathbf{g}\, e^{i(k+2\pi g)\cdot r}$, where a_0 and a_g are given by 6.24. Writing $a_g^\dagger = a_g\, e^{-i\delta}$, we can express the sum of these two waves in the form

$$a_0\, e^{ik\cdot r} + a_g e^{i(k+2\pi g)\cdot r}$$

$$= [(a_g^\dagger + a_0)\cos(\pi g.r + \delta/2) + i(a_g^\dagger - a_0)\sin(\pi g.r + \delta/2)] \times e^{i\{(k+\pi g).r+\delta/2\}}$$

$$= [a_0^2 + a_g^2 + 2a_0\, a_g^\dagger \cos(2\pi g.r + \delta)]^{\frac{1}{2}}\, e^{i\{(k+\pi g).r+\delta/2+\eta\}} \qquad 6.53$$

where
$$\tan\eta = \frac{a_g^\dagger - a_0}{a_g + a_0}\tan(\pi g.r + \delta/2).$$

If the wave vector **k** of the first component wave is in exactly the Bragg direction KO of Fig. 6.2(b), then the wave vector $\mathbf{k} + \pi\mathbf{g}$ appearing in the exponential factor in 6.53 represents a wave travelling *parallel to the reflecting planes of atoms*. The direction of propagation of the wave 6.53 will remain nearly parallel to the reflecting planes of atoms provided the wave vector k is near enough to the Bragg direction for the component wave $a_g\, e^{i(\mathbf{k}+2\pi\mathbf{g})\cdot\mathbf{r}}$ to have appreciable amplitude.

The amplitude of the Bloch wave 6.53 is not constant across the wave front, but varies as indicated by the term in square brackets. Using 6.24, we may write this amplitude factor as

$$\tfrac{1}{2}\left[1 + \frac{W}{\sqrt{(1+W^2)}}\right]\left[1 - \frac{1}{\sqrt{(1+W^2)}}\cos(2\pi\mathbf{g}\cdot\mathbf{r} + \delta)\right]^{\frac{1}{2}} \qquad 6.54a$$

This clearly has the periodicity $1/g$ of the reflecting planes, and has a *maximum* when $2\pi\mathbf{g}.\mathbf{r} + \delta = (2n+1)\pi$, where n is an integer. To see the significance of this, we recall that the Fourier coefficient of the

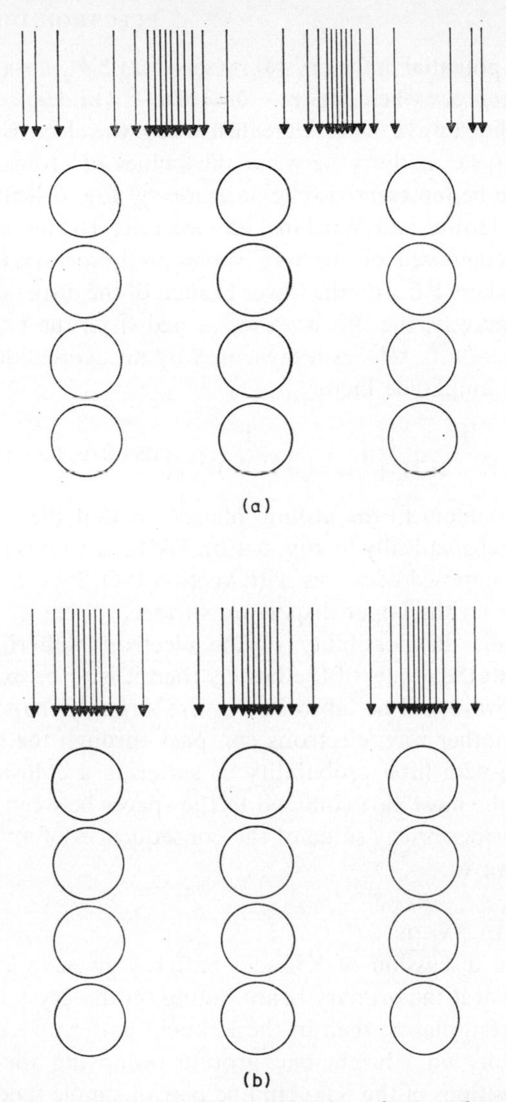

Fig. 6.4. Bloch waves formed by the combination of a crystal wave corresponding to the incident beam and one corresponding to the diffracted beam. The density of the arrows representing the wave vectors is proportional to the amplitude of the waves. The Type I wave shown in (a) is confined largely to the spaces between the atomic planes, and therefore suffers little absorption. The Type II wave shown in (b) travels mainly along the lines of atoms, and is therefore strongly absorbed.

electrostatic potential in the crystal is (see § 6.2) $2|V_g|\cos(2\pi_g.\mathbf{r}+\delta)$ and has a *minimum* when $2\pi\mathbf{g}\cdot\mathbf{r}+\delta=(2n+1)\pi$. Hence the amplitude of the Bloch wave 6.45a is greatest where the electrostatic potential is lowest, i.e. midway between the planes of atoms. The wave can therefore be represented schematically by Fig. 6.4(a). Following Hashimoto, Howie and Whelan [30], we refer to this as a Type I wave. It is composed of the two waves with vectors PO, PG, in Fig. 6.2(b), where P lies on the lower branch of the dispersion surface.

In the same way, the Bloch wave formed from the pair of waves $a_0'\,e^{i\mathbf{k}\cdot\mathbf{r}}$ and $a_g'\,e^{i(\mathbf{k}+2\pi\mathbf{g})\cdot\mathbf{r}}$ is represented by an expression similar to 6.45 with an amplitude factor

$$\tfrac{1}{2}\left[1-\frac{W}{\sqrt{(1+W^2)}}\right]\left[1+\frac{1}{\sqrt{(1+W^2)}}\cos(2\pi\mathbf{g}.\mathbf{r}+\delta)\right]^{\frac{1}{2}}. \quad 6.54\text{b}$$

This is a maximum in the atomic planes, so that the wave can be represented schematically by Fig. 6.4(b). We refer to this as a Type II wave: it is composed of waves with vectors P′O, P′G in Fig. 6.2(b), where P′ lies on the upper dispersion surface.

If we admit the possibility of the electrons suffering inelastic collisions with the atoms of the lattice, then it is to be expected that the Type II wave will be ‘absorbed’ more strongly than the Type I wave. Put another way, electrons can pass through the crystal as a Type I wave with little probability of suffering a collision because they are for the most part confined to the spaces between atoms. We will now consider briefly some of the consequences of an absorption of Type II waves.

6.5.1 KIKUCHI BANDS

In the simple discussion of Kikuchi patterns given in § 4.6, it was mentioned that if the primary beam falling on the crystal is parallel to a set of atom planes, then in the Kikuchi pattern we find a band of low intensity on a bright background occupying the space between the positions of the Kikuchi line pair of simple theory. We are now in a position to explain this effect. Diffuse scattering of the primary beam will give rise to a multitude of Type I and Type II Bloch waves corresponding in Fig. 6.2(b) to pairs of waves associated with a multitude of points P and P′ distributed over the dispersion surface. If the crystal is sufficiently thick, then the Type II waves will be largely absorbed and only Type I waves will emerge from the far side of the crystal. Let us examine these waves further. In Fig. 6.2(b), let us suppose that the point P lies to the left of K, as indicated in the figure. Then the wave PO of amplitude a_0 makes a glancing angle with the atom planes (which are perpendicular to OG) which is greater than the Bragg angle $\frac{1}{2}$OKG. The wave PG, of amplitude a_g, on the other hand, has a glancing angle less than the Bragg angle. The wave PO will therefore contribute to the part of the Kikuchi pattern lying *outside* the Kikuchi band. The wave PG will contribute to the band itself. But we have defined the quantity W to be positive when P lies to the right of K, and under these conditions we see from equation 6.24 that $a_0 > a_g$. Hence the portion of the Kikuchi pattern inside the band is less intense than the portion outside. A similar argument applied to the case when P lies to the left of K, and W is then negative, leads to the same final result.

Experimentally, it is found that although for sufficiently thick crystals the Kikuchi band always has a lower intensity than the surrounding background, thin crystals may give rise to Kikuchi bands of *enhanced* intensity, and at an intermediate thickness of crystal the band disappears. To explain this, we have to assume that the diffuse scattering processes initially generate mainly Type II waves. This is reasonable since the diffuse scattering of electrons will occur at atomic sites, and therefore the scattered wave will have its largest amplitude at these sites. These may subsequently undergo further scattering and produce some Type I waves, which will be transmitted through the crystal with relatively little absorption. In thin crystals, therefore, Type II waves will predominate while in thick crystals Type I waves are dominant.

6.5.2 VISIBILITY OF EXTINCTION CONTOURS

We have noted in §§ 3.5 and 6.3 that the intensities of the primary and diffracted waves transmitted through a crystal lamina are periodic functions of the thickness of the lamina; and in § 10.3 we note that electron micrographs of crystals frequently show light and dark lines, which are fringes of constant thickness. These fringes arise from interference between a Type I wave (consisting of component waves represented in Fig. 6.2(b) by the wave vectors PO and PG) and the corresponding Type III wave (consisting of component waves represented by P′O and P′G). It is clear that interference effects will not be observed in a thick crystal since the Type II wave is then completely absorbed. In fact, the usual way of measuring absorption is to observe the decrease in visibility of these fringes with increasing crystal thickness.

6.6 *Kikuchi envelopes*

Reference has been made in § 4.6, Figs. 4.9 and 4.10, to the distortion of Kikuchi lines near positions where two lines intersect. These distortions are due to dynamical interactions between beams reflected at two sets of atomic planes. Although the exact details of the process are still far from clear, it is possible to give a qualitative description of the phenomenon in terms of the dynamical theory.

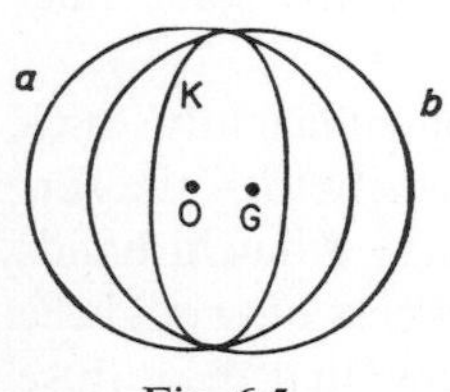

Fig. 6.5.

We first recall that the simple kinematic theory states that a strong diffracted beam will occur if a point L – the Laue point – can be found in reciprocal space which is distant $1/\lambda$ both from the origin O and from a reciprocal lattice point G (Fig. 3.4(b)). The directions of the diffracted beams are therefore formed by drawing a sphere *a* of radius $1/\lambda$ centred on O of the reciprocal lattice (Fig. 6.5). The intersection K of this sphere with a similar sphere *b* centred on a lattice point G is the locus of the Laue point.

A little consideration of the simple theory given in § 4.6 will show that a Kikuchi line is formed by the projection of part of the locus K from the point O on to the photographic plate. Projection of the same locus from the point G gives the position of the other line of the pair.

The spheres *a* and *b* of Fig. 6.5 are referred to as the *dispersion surfaces* of the primary and diffracted waves. The dynamical theory (§ 6.2, Fig. 6.1) states that near their intersection, the dispersion surfaces *a* and *b* (the notation is the same in Figs. 6.1 and 6.5) are modified so as to become the branches + and − of an hyperbola.

When we come to consider the intersection of two Kikuchi lines, we obviously have to deal with diffracted waves corresponding to *two* reciprocal lattice points G and H. If we use the simple kinematic theory, we shall draw dispersion surfaces in the form of spheres of radii $1/\lambda$ centred on the origin O and two points G and H of the reciprocal lattice. One Kikuchi line is the projection from O of the intersection of the spheres centred on O and G; the other is the projection from O of the intersection of the spheres centred on O and H. The point where the two Kikuchi lines intersect is the projection of the point where all three spheres intersect.

In terms of the dynamical theory, we may represent the intersection of the two spheres G and H as in Fig. 6.6. *a* and *b* are portions of the spheres; their radius of curvature is so large that over the whole of the figure they are indistinguishable from planes. The dynamical theory replaces these two intersecting planes by two

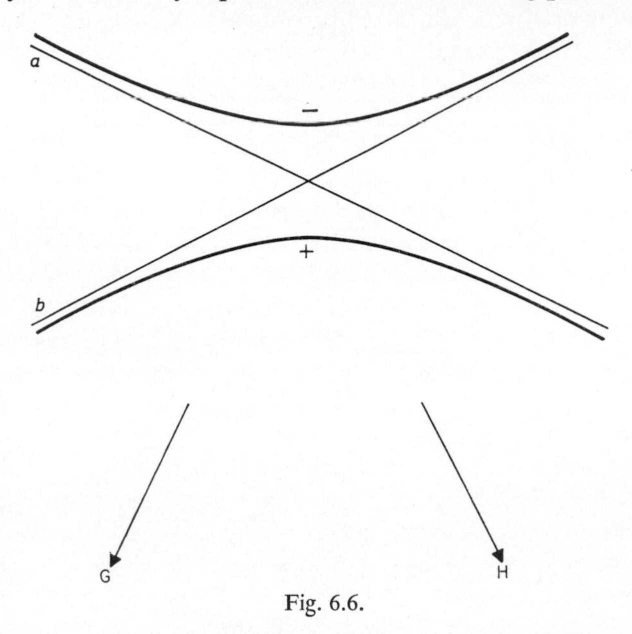

Fig. 6.6.

sheets of a hyperbolic cylinder + and −. (Fig. 6.6 is, of course, exactly the same as Fig. 6.1 except that the points O and G of the latter figure are now the points G and H.) The origin O of the reciprocal lattice will in general be outside the plane of Fig. 6.6, and the spherical dispersion surface about it will approximate to a plane inclined to the plane of the figure. Far from the centre of the figure, this plane will intersect the asymptotic planes *a*, *b* of the hyperbolic cylinder in straight lines. The projection of these from O on to the photographic plate will correspond to the outer parts of the Kikuchi lines *aa'*, *bb'* represented in Fig. 4.9. Near the centre of the figure, the same plane intersects the hyperbolic cylinder in a hyperbola, and the projection of this from O on to the photographic plate will correspond to the curves *ab'* and *ba'* of Fig. 4.9. The dynamical theory thus gives a good qualitative description of the main features of Kikuchi line intersections, and therefore of Kikuchi envelopes.

7

Low Energy Electron Diffraction

7.1 *Introduction*

The early work of Davisson and Germer on low energy electron diffraction (nowadays commonly referred to as LEED) has been discussed in § 1.2, and present-day LEED cameras are described in §§ 2.10, 2.11. In this chapter, we shall indicate the nature of the results obtained with such instruments.

Since low energy electrons can penetrate only a few layers of atoms near the surface of a crystal, LEED phenomena will be profoundly influenced by the nature of the surface of the crystal, and, in particular, by the presence of adsorbed gas. The main use of LEED is therefore for studying surface properties. It is obviously desirable, when possible, to examine the structure of a clean, gas-free surface before studying the effects of gas contamination.

7.2 *Clean surfaces: metals*

The most useful LEED observations are those made on low index surfaces of single crystals, since only for such surfaces will there be a well-defined pattern of surface atoms. In the case of metals, such surfaces are produced by cutting the specimen in the appropriate orientation from a single crystal and polishing the surface mechanically with successively finer abrasives. This treatment is followed by electrolytic polishing to remove the surface layer of metal which has been deformed by the mechanical polishing.

The next problem is to remove any adsorbed gas from the surface of the crystal. Each metal requires a different treatment, which can be discovered only by a process of trial and error. Some metals, such as tungsten, can be cleaned simply by heating to a high temperature in an ultra-high vacuum ($\sim 10^{-10}$ torr). In the case of nickel, on the other hand, it is necessary to heat the metal in an atmosphere first of oxygen at a pressure of 10^{-6} torr and then of hydrogen at a pressure of about 10^{-7} torr before finally heating in ultra-high vacuum to remove the hydrogen (Germer and MacRae [25]). Another useful method is to bombard the surface with argon ions

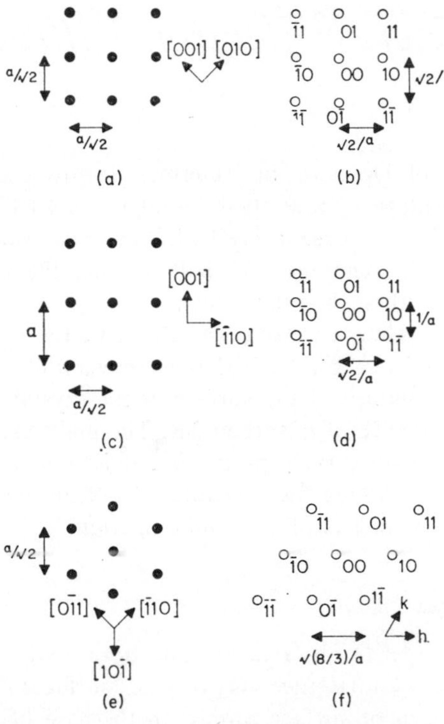

Fig. 7.1. LEED patterns from face-centred cubic crystal. The solid discs represent atoms in the surface of the crystal, the open circles represent reciprocal lattice points corresponding to these surface planes of atoms.

(a) Arrangement of atoms in (100) face.
(b) Reciprocal lattice corresponding to (a).
(c) Arrangement of atoms in (110) face.
(d) Reciprocal lattice corresponding to (c).
(e) Arrangement of atoms in (111) face.
(f) Reciprocal lattice corresponding to (e).

having an energy of a few hundred eV (Farnsworth, Schlier, George and Burger [23]).

As examples of the type of diffraction pattern to be expected, let us consider the patterns produced by the (100), (110), and (111) faces of a face-centred cubic crystal such as nickel. Figure 7.1(a), (c) and (e) show the arrangement of atoms on these surfaces, assuming that the surface layer of atoms has the same structure as that of the bulk material. This two-dimensional arrangement of atoms produces a diffraction pattern which can be indexed in terms of a two-dimensional reciprocal net shown in Fig. 7.1(b), (d) and (f). The diffraction pattern observed on the hemispherical fluorescent screen of a LEED camera of the type described in § 2.10 is a distorted representation of such a net. The distance of a spot from the centre of the pattern gives a measure of the angle θ between the incident ray falling on the crystal and the diffracted ray. This is related to the distance D_{hk}

TABLE 2. Distance D_{hk} between rows of atoms in the surface of a face-centred cubic crystal corresponding to a point *hk* of the reciprocal lattice shown in Fig. 7.1.

Miller indices of crystal face	D_{hk}
(100)	$a/[2(h^2 + k^2)]^{\frac{1}{2}}$
(110)	$a/(2h^2 + k^2)^{\frac{1}{2}}$
(111)	$a[3/8(h^2 + hk + k^2)]^{\frac{1}{2}}$

between lines of atoms in the surface of the crystal by equation 1.5, which may be re-written

$$D_{hk} \sin \theta = \lambda, \qquad 7.1$$

where D_{hk} is given in Table 2.

If the potential on the electron gun is increased, thus decreasing the electron wavelength λ, then, as may be seen from equation 7.1, the spots of the diffraction pattern move inwards towards the centre of the pattern. It is observed that the intensities of the spots do not remain constant. Figure 7.2 which is based on observations of Park and Farnsworth [56] on the diffraction pattern from a (111) surface of nickel, shows that for each spot there are certain critical voltages which produce a maximum intensity. This is due to interference between the waves scattered by successive layers of atoms.

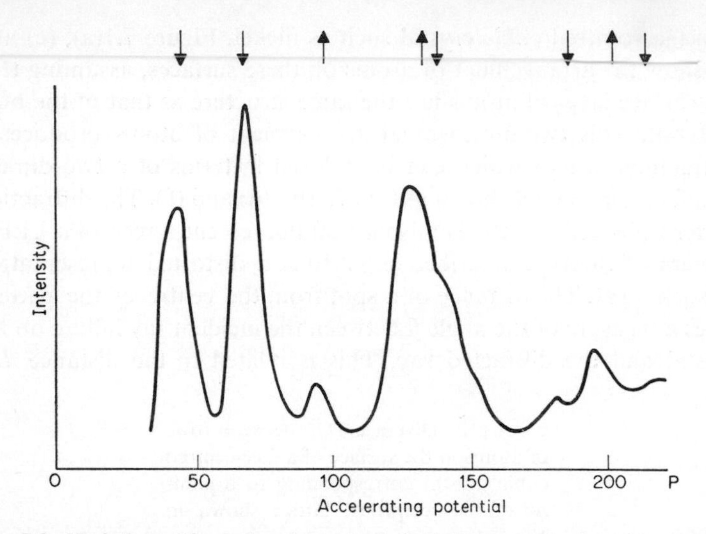

Fig. 7.2. Variation with accelerating voltage P of the intensity of the 10 reflection from the 111 face of nickel. The arrows indicate the position of the intensity maxima calculated from equation 7.3. The arrows pointing downwards correspond to case (a), and those pointing upwards to case (b).

In Fig. 7.3, ACFD is the surface layer of atoms of a crystal, A and F being adjacent rows of atoms in the surface at a distance D_{hk} apart. The angle θ between the incident and diffracted waves is given by equation 7.1. BG is the next layer of atoms at a depth d below the surface of the crystal. The atom rows in this layer will be

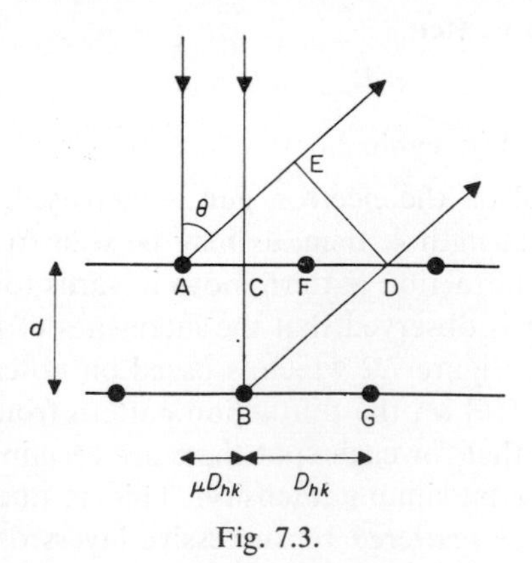

Fig. 7.3.

displaced as shown in the figure a distance μD with respect to those in the surface layer. The fraction μ can be found for any given diffraction by considering the arrangement of the atoms in the complete, three-dimensional, crystal. For example, μ is $\frac{1}{2}$ for the 10 diffraction from the 100 surface and 0 for the 11 diffraction from the same surface. From the figure, the path difference between waves scattered by atoms A and B in the two layers is

$$\begin{aligned} \mathrm{CB} + \mathrm{BD} - \mathrm{AE} &= d + d \sec \theta - (\mu D + d \tan \theta) \sin \theta \\ &= d(1 + \cos \theta) - \mu D \sin \theta. \end{aligned}$$

For constructive interference, this must be an integral number n of wavelengths. Then, making use of 7.1 we obtain

$$d(1 + \cos \theta) = (n + \mu)\lambda. \qquad 7.2$$

Eliminating θ between 7.1 and 7.2 and making use of the relation 1.3a between the voltage P on the electron gun and the wavelength λ we obtain the following expression for the voltage at an intensity maximum

$$P = 3{\cdot}75 \times 10^5 \left[\frac{(n + \mu)^2 + d^2/D_{hk}{}^2}{d(n + \mu)}\right]^2 \text{ volts} \qquad 7.3$$

(d and D_{hk} in pm).

In deriving equation 7.3, we have neglected the effect of the inner potential of the crystal. There are three cases to be considered.

(*a*) If there is on the surface of the crystal a step of height equal to one atomic layer, then the condition for constructive interference of the beams diffracted from the portions of the surface on either side of the step is correctly given by 7.3. For since neither beam penetrates below the surface of the crystal, the inner potential can have no effect.

(*b*) The condition for constructive interference between waves diffracted by successive planes of atoms *within* the crystal is obviously given by writing $(P + V_0)$ where V_0 is the inner potential, in place of P in 7.3.

(*c*) The condition for constructive interference between waves diffracted by the surface layer and the next layer, when account is taken of the inner potential, is (see Fig. 7.3)

$$\frac{\mathrm{CB} + \mathrm{BD}}{\lambda_1} - \frac{\mathrm{AE}}{\lambda_2} = n,$$

where λ_1 and λ_2 are respectively the wavelengths of the electrons in the crystal and in vacuum. This condition leads to an expression for the voltage at an intensity maximum which is intermediate between those appropriate to cases (*a*) and (*b*).

In Fig. 7.2, the calculated positions of the maxima in cases (*a*) and (*b*) are marked with arrows, and it will be observed that both maxima occur in the curves. The lack of resolution of the maxima may perhaps be due in part to some interference of type (*c*) above.

The intensity maxima are calculated from 7.3 using the quantities

$$\mu = \tfrac{1}{3} \qquad D_{10} = (\tfrac{3}{8})^{\frac{1}{2}}a \text{ (see Table 7.1)}$$
$$d = a/\sqrt{3} \qquad a = 352 \text{ pm}$$
$$\text{Inner potential } V_0 = 16 \text{ V}.$$

There are indications of other weak maxima in Fig. 7.2. Germer, MacRae and Hartman [26] have claimed that these can be fitted into an equation of the type of 7.3 by assuming a value for d about 5% less than the interplanar spacing of the bulk crystal. Park and Farnsworth [55], on the other hand, maintain that these maxima are due to stacking faults in the crystal. These would change the value of μ in 7.3. The impossibility of determining the precise position of these weak maxima makes it difficult to distinguish between these alternatives.

7.3 *Clean surfaces: non-metals*

The diffraction patterns from most clean metal surfaces can be interpreted as indicated in § 7.2 assuming the surface layer of atoms to have the same structure as the bulk crystal. Crystals in which the atoms have covalent bonds behave differently. Diffraction patterns from the surfaces of such crystals are characterized by the occurrence of spots with fractional indices, if these are indexed on the assumption that the surface atoms have the same arrangement as in the bulk crystal. This implies that the surface atoms are arranged in a mesh which is a multiple of that of parallel layers in the bulk crystal. It is possible by a careful study of the relative intensities of the various diffractions to get an indication of the detailed atomic arrangement using the methods of three-dimensional crystallography. Thus Lander and Morrison [40] have proposed the arrangements shown in Fig. 7.4 for the atoms in the 100 and 111 surfaces of silicon and germanium crystals. (It will be observed that in each case the density of atoms in the surface layer is less than in the bulk.) This rearrangement of the surface atoms is readily accounted for: atoms on the surface have 'dangling bonds', and these will bind the atoms together in groups, thus displacing them from their normal positions.

Various difficulties arise in the determination of these surface structures. In the first place, different methods of preparing the specimen often produce different structures. Secondly, the number of parameters involved is large. The atoms may be displaced not merely in the surface layer as suggested in Fig. 7.4, but also in the perpendicular direction; and the atoms in a few layers near the surface

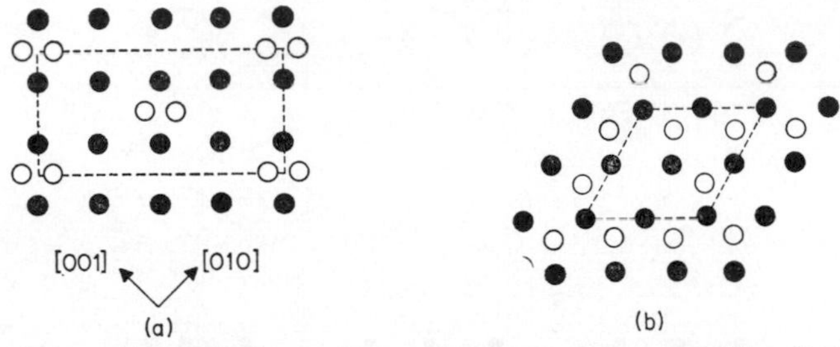

Fig. 7.4(a). Arrangement of atoms in 100 planes of Ge or Si. The discs show the arrangement of atoms in planes in the interior of the crystal; the open circles show the arrangement of atoms in the surface layer. Note the large centred mesh indicated by dotted lines.

(b) Arrangement of atoms in 111 planes of Ge or Si. The discs show the arrangement of atoms in planes in the interior of the crystal; the open circles show the arrangement of atoms in the surface layer. Note the large mesh indicated by dotted lines.

will be expected to have arrangements intermediate between that of the surface layer and that of the bulk crystal. Yet a further complication arises because a large phase shift occurs when low energy electrons are scattered by atoms (Mott and Massey [53]). This phase shift is strongly dependent on the energy of the electrons and also on the structure of the atoms. It does not affect the diffraction pattern of a crystal of an *element*. Crystals of chemical *compounds* on the other hand, will have more than one atomic species in the surface layer, and these will have phase shifts differing by an unknown amount. The effect on the diffraction pattern of such a phase difference between groups of atoms of different species will be almost indistinguishable from the effect of a relative displacement of the two groups of atoms. Because of these various effects, too much reliance must not be placed on the finer details of any proposed structure.

7.4 *Diffraction patterns due to adsorbed gas*

In order to show the kind of information on gas adsorption which can be provided by LEED, we shall review some of the work of Germer and Macrae [25] on the adsorption of oxygen on nickel.

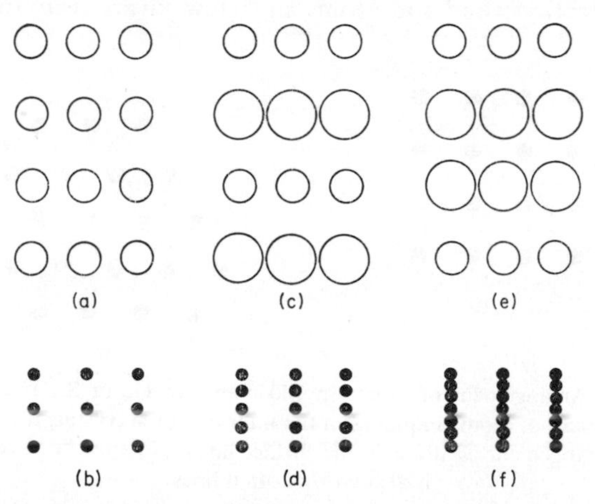

Fig. 7.5. Structure of clean nickel 110 surface before and after exposure to oxygen. The small circles indicate the positions of nickel atoms and the large ones the positions of nickel atoms covered by oxygen atoms: the latter have a lower scattering power. The dots indicate points of the reciprocal lattice, and therefore the appearance of the diffraction pattern.

(a) Clean nickel surface.
(b) Reciprocal net corresponding to (a).
(c) Nickel surface having half the nickel atoms covered by oxygen atoms.
(d) Reciprocal net corresponding to (c).
(e) Nickel surface having $\frac{2}{3}$ of the nickel atoms covered by oxygen atoms.
(f) Reciprocal net corresponding to (e).

Figure 7.5(a) shows the surface layer of atoms on a clean 110 surface of a nickel crystal. Figure 7.5(b) shows the reciprocal net corresponding to this; this reciprocal net can also be regarded as a schematic representation of the diffraction pattern from such a surface. If oxygen at a pressure of 5×10^{-9} torr is admitted to the camera, the spots of the diffraction pattern become joined by vertical streaks. After about three minutes, these streaks have shortened and coalesced into spots lying in the positions h, k with k half integral. The diffraction pattern now appears as in Fig. 7.5(d). These new spots have exactly the same intensity as the original ones of integral h, k. In other words Fig. 7.5(d) is the reciprocal net of a new surface structure of nickel atoms shown in Fig. 7.5(c). A reconstruction of the surface has taken place resulting in the removal of alternate rows of nickel atoms.

If it be assumed that every oxygen molecule striking the surface dissociates into two atoms which remain attached to the surface, then an exposure of three minutes at an oxygen pressure of 5×10^{-9} torr would just suffice to place one oxygen atom in each unit mesh of the structure of 7.5(c). One may therefore reasonably conjecture that the oxygen atoms are located as shown by the large circles in Fig. 7.5(c). It is not possible to test this conjecture, because the electron scattering power of oxygen is much less than that of nickel, so that the oxygen atoms have negligible effect on the intensities of the spots in the diffraction pattern.

If a 110 nickel surface which has been exposed to oxygen as described above is given further oxygen treatment ($\frac{1}{2}$ hr at 10^{-8} torr, or proportionally less at higher pressure), the pattern changes again to that of Fig. 7.5(f), corresponding to the surface structure of nickel atoms shown in Fig. 7.5(c). If it be assumed that each mesh of this structure contains two oxygen atoms, then the length of time taken for the pattern to develop indicates that the sticking probability of oxygen atoms on to the structure of Fig. 7.5(c) is only 5%. Clearly, the second group of atoms are much less tightly bound to the surface than the first. It is therefore not surprising to find that the structure of Fig. 7.5(e) reverts to that of Fig. 7.5(c) on heating in high vacuum to 620°K; though the latter structure does not go over to that of a clean surface (Fig. 7.5(a)) on heating to the highest temperatures.

8

Electron Diffraction by Gases

8.1 *Introduction*

The study of the structure of molecular gases is one of the most successful applications of electron diffraction. We have seen in other chapters that electron diffraction has been used for the determination of the structure of some crystals. However, the vast majority of known crystal structures have been determined by X-ray diffraction. In the case of gases, on the other hand, although some early work was carried out using X-rays, almost all our detailed knowledge of their molecular structure has been obtained by electron diffraction. There are two reasons for this. First, the total number of molecules in a gaseous sample is necessarily small, so that it is essential to make use of the greater scattering cross-section for electrons in order to obtain a satisfactory diffraction pattern in a reasonable time. The second, and much more important, reason is the following. It will be seen in § 8.2 that the main features of the diffraction pattern are determined by a factor

$$\sin\left[\frac{4\pi}{\lambda} l \sin \tfrac{1}{2}\phi\right] \tag{8.1}$$

where l is the distance between two atoms in the molecule and ϕ is the angle between the incident and diffracted rays. To obtain an accurate value of l, it is desirable to observe as many diffraction rings (maxima of 8.1) as possible. Since the maximum range of $\sin \frac{1}{2}\phi$ is from 0 to 1, the maximum number of rings which can be observed is about $4l/\lambda$. It is therefore advantageous to use electrons rather than X-rays, which have a wavelength more than an order of magnitude greater.

8.2 *Simple theory of diffraction by gas molecules*

We first consider the simplest case of a diatomic molecule consisting of two atoms i and j separated by a vector distance $\mathbf{l}_{ij}$. The atom i

will scatter a spherical wave of the form given by 3.1 and 3.2:

$$\frac{-e^{ikr}}{r}\frac{me^2\lambda^2(Z_i - F_i)}{8h^2\pi\varepsilon_0 \sin^2 \frac{1}{2}\phi}. \qquad 8.2$$

Since we are dealing with a gas, it is more convenient to write this expression in terms of the angle ϕ between the incident and diffracted beams instead of the Bragg angle $\theta = \frac{1}{2}\phi$ which is used in other parts of this book. Z_i and F_i are, of course, the atomic number and atomic scattering factor of atom i.

The phase difference between the waves scattered by the atoms i and j is (see equation 3.5 and Fig. 3.2)

$$-k\mathbf{l}_{ij}\cdot(\mathbf{s} - \mathbf{s}_0).$$

The total amplitude scattered by the molecule is therefore

$$\psi = \frac{-e^{ikr}}{r}\frac{me^2\lambda^2}{8h^2\pi\varepsilon_0 \sin^2 \frac{1}{2}\phi}[(Z_i - F_i) + (Z_j - F_j)\exp\{-k\mathbf{l}_{ij}(\mathbf{s} - \mathbf{s}_0)\}]$$

and the scattered intensity is

$$\psi\psi^* = \frac{m^2e^4\lambda^4}{64h^4r^2\pi^2\varepsilon^2_0 \sin^4 \frac{1}{2}\phi}[(Z_j - F_i)^2 + (Z_j - F_j)^2 + 2(Z_i - F_i)(Z_j - F_j)\cos k\mathbf{l}_{ij}\cdot(\mathbf{s} - \mathbf{s}_0)]. \qquad 8.3$$

Following the argument in Chapter 3, p. 54, we see that the scattering cross-section σ of the molecule is given by 8.3 with the factor r^2 omitted from the denominator.

The first two terms of 8.3 correspond to scattering by the individual atoms. The last term corresponds to the scattering by the whole molecule, and contains the interference term $\cos k\mathbf{l}_{ij}\cdot(\mathbf{s} - \mathbf{s}_0)$ which describes the effect of the molecular structure on the diffraction pattern.

The expression 8.3 gives the total intensity of the electrons scattered *without change of energy*. Some electrons are scattered inelastically, giving up part of their energy in exciting the atoms comprising the molecule. This inelastic scattering is considered to be the sum of the effects due to the individual atoms in the molecule (scattering accompanied by excitation of *molecular* vibrations or rotations is ignored). According to Morse [52] the effect of this inelastic scattering is to add to the expression in the square bracket in 8.3 terms

$Z_i q_i + Z_j q_j$. The quantities q have been tabulated by Bewilogua [6]. It is sufficient for our present purpose to note that q is zero for $\phi = 0$, and rises asymptotically to unity for large ϕ. In practice, q differs little from unity except near the centre of the diffraction pattern.

A gas, of course, contains molecules in all orientations, and we must therefore average the cosine factor in 8.3 over all orientations of the molecular axis $\mathbf{l}_{ij}$ with respect to the scattering vector $(\mathbf{s} - \mathbf{s}_0)$. We first note that since $\mathbf{s}$ and $\mathbf{s}_0$ are unit vectors inclined to each other at an angle ϕ, the magnitude

$$|\mathbf{s} - \mathbf{s}_0| = 2 \sin \tfrac{1}{2}\phi.$$

Hence, if α is the angle between the vectors $\mathbf{l}_{ij}$ and $(\mathbf{s} - \mathbf{s}_0)$, the argument of the cosine factor is $2kl_{ij} \sin \frac{1}{2}\phi \cos \alpha$. If β is the azimuth of the vector $\mathbf{l}_{ij}$ about $(\mathbf{s} - \mathbf{s}_0)$, then the probability that $\mathbf{l}_{ij}$ lies in the element of solid angle

$$d\omega = \sin \alpha \, d\alpha \, d\beta$$

is

$$\frac{1}{4\pi} \sin \alpha \, d\alpha \, d\beta$$

Hence the mean value of the cosine factor is

$$\langle \cos (2k\mathbf{l}_{ij} \sin \tfrac{1}{2}\phi \cos \alpha) \rangle = \int_0^{2\pi} \int_0^{\pi} \cos (2k\,\mathbf{l}_{ij} \sin \tfrac{1}{2}\phi \cos \alpha) \frac{\sin \alpha}{4\pi} d\alpha \, d\beta$$

$$= \frac{\sin \mathscr{S} l_{ij}}{\mathscr{S} l_{ij}} \qquad 8.4$$

where

$$\mathscr{S} = 2k \sin \tfrac{1}{2}\phi = \frac{4\pi}{\lambda} \sin \tfrac{1}{2}\phi. \qquad 8.5$$

The expression 8.3 for the intensity scattered by a diatomic molecule can obviously be generalized to the case of a polyatomic molecule. Introducing 8.4 and 8.5, and including the terms to represent the inelastic scattering, we see that the scattering cross section of a polyatomic molecule can be written as

$$\sigma = \frac{4\pi^2 m^2 e^4}{h^4 \varepsilon_0{}^2} \frac{1}{\mathscr{S}^4} \Big[\Sigma (Z_i - F_i)^2 + \Sigma Z_i q_i$$

$$+ \sum_{ij}{}' (Z_i - F_i)(Z_j - F_j) \frac{\sin \mathscr{S} l_{ij}}{\mathscr{S} l_{ij}} \Big] \qquad 8.6$$

where the prime on the third summation sign indicates that terms for which $i = i$ are to be omitted.

8.3 *Effect of molecular vibrations*

In deriving equation 8.6, it has been tacitly assumed that the interatomic distances l_{ij} are *constant*, i.e. that the molecule behaves as a rigid body. In fact, the atoms will be vibrating about their equilibrium positions. The period of these vibrations is many orders of magnitude greater than the time during which an electron interacts with the molecule. Consequently, in considering the diffraction process, it is correct to regard the molecule as a rigid body, and thus to derive 8.6. In this expression, however, the quantities l_{ij} are fluctuating with the frequency of the molecular vibrations. Our problem is therefore to calculate the average values of terms of the form $(\sin \mathscr{S}l)\mathscr{S}l$, where for convenience we have dropped the suffices i, j.

Assuming that the atoms execute harmonic oscillations, the instantaneous value of an interatomic distance may be written as $l + \delta$, where l is its equilibrium value. The probability of δ lying in the range δ to $\delta + d\delta$ is

$$\frac{\Delta}{\sqrt{\pi}} \exp \{-\delta^2/\Delta^2\}\, d\delta.$$

Δ is the root mean square value of the fluctuation of l from its equilibrium value; it is typically a few per cent of l, and is mainly due to the zero-point energy of the molecular vibration. Hence the average value of $(\sin \mathscr{S}l)/\mathscr{S}l$ is

$$\langle(\sin \mathscr{S}l)/\mathscr{S}l\rangle = \int_{-\infty}^{+\infty} \frac{\sin \mathscr{S}(l+\delta)}{\mathscr{S}(l+\delta)} \frac{\Delta}{\sqrt{\pi}} \exp \{-\delta^2/\Delta^2\}\, d\delta$$

We may without serious error omit the term δ in the denominator of the integrand, since the exponential factor ensures that the only appreciable contribution to the integral comes from the range where δ is a few per cent of l. Thus

$$\langle(\sin \mathscr{S}l)/\mathscr{S}l\rangle = \frac{\Delta}{\mathscr{S}l\sqrt{\pi}} \int_{-\infty}^{+\infty} (\sin \mathscr{S}l \cos \mathscr{S}\delta + \cos \mathscr{S}l \sin \mathscr{S}\delta) \exp \{-\delta^2/\Delta^2\}\, d\delta$$

The second term of the integrand, being an odd function of δ, contributes nothing to the integral. Thus, on writing $\cos \mathscr{S}\delta$ in exponential form by means of Euler's identity we get

$$\langle(\sin \mathscr{S}l)/\mathscr{S}l\rangle = \frac{\Delta \sin \mathscr{S}l}{2\sqrt{(\pi)}\mathscr{S}l} \int_{-\infty}^{+\infty} \left[\exp\left\{ -\left(\frac{\delta}{\Delta} + i\frac{\mathscr{S}\Delta}{2}\right)^2 - \frac{\mathscr{S}^2\Delta^2}{4} \right\} + \text{complex conjugate} \right]$$

$$= \exp\{-\mathscr{S}^2\Delta^2/4\} \frac{\sin \mathscr{S}l}{\mathscr{S}l}. \qquad 8.7$$

The effect of the molecular vibrations is therefore to modify equation 8.6 to

$$\sigma = \frac{4\pi^2 m^2 e^4}{h^4 \varepsilon_0{}^2 \mathscr{S}^4} \left[\Sigma (Z_i - F_i)^2 + \Sigma Z_i q_i + \sum_{ij}{}'(Z_i - F_i)(Z_j - F_j) \frac{\sin \mathscr{S}l_{ij}}{\mathscr{S}l_{ij}} \exp\{-\mathscr{S}^2\Delta^2{}_{ij}/4\} \right]. \qquad 8.8$$

8.4 *Interpretation of diffraction patterns of molecular gases*

The apparatus used for producing diffraction patterns from gases has been described in § 2.12. In considering how equation 8.8 may be used to interpret these patterns, we first note that in the outer parts of the pattern, where $\mathscr{S}$ and the angle of diffraction ϕ are fairly large, $F \to 0$ and $q \to 1$. Thus 8.8 approximates to

$$\sigma = \text{const.} \frac{1}{\mathscr{S}^4} \left[\Sigma (Z_i{}^2 + Z_i) + \sum_{ij}{}' Z_i Z_j \frac{\sin \mathscr{S}l_{ij}}{\mathscr{S}l_{ij}} \exp\{-\mathscr{S}^2\Delta^2{}_{ij}/4\} \right]. \qquad 8.8'$$

The quantity $\mathscr{S}$ is given in equation 8.5. With the usual approximation $\sin \phi \approx \phi \approx R/L$ where R is the radius of a diffraction ring and L is the camera length

$$\mathscr{S} = 2\pi R/\lambda L. \qquad 8.5'$$

The $\mathscr{S}^{-4}$ factor in 8.8′ shows that there is a *very rapid* diminution of intensity as one moves radially outwards in the pattern; in practice, this amounts to a ratio of at least a million between the intensities at the centre and at the edge of the pattern. This, of course ranges far beyond the linear portion of the density/exposure part of the characteristic curve of the photographic plate. It is for this reason that, as described in § 2.12, a rotating sector device is used which has the effect of giving the outer parts of the pattern a longer exposure than the inner portion. Consideration of 8.8′ and 8.5′ would suggest that the section should be shaped so as to give an exposure proportional to R^4 or R^5. In fact, the effect of the term F in 8.8 is to reduce somewhat the intensity near the centre of the pattern so that in practice it is found that a sector giving an exposure proportional to R^3 gives the most uniform blackening of the photographic plate.

In equation 8.8′, the first summation, representing the scattering by the individual atoms, is much greater than the second term, which

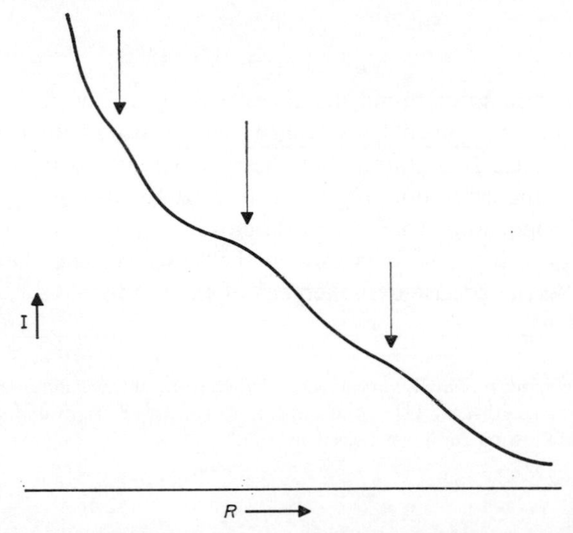

Fig. 8.1. Sketch of typical microphotometer trace of a diffraction pattern of a gas. The positions of diffraction maxima are indicated by arrows. These diffraction maxima produce only slight changes in the gradient of the monotonically decreasing background intensity.

is due to the molecular structure. Hence the general appearance of a microphotometer trace of a diffraction pattern, even when the rotating sector device is used, is that of a monotonically falling curve; the effect of the second term in equation 8.8′ is simply to produce changes of slope as indicated schematically in Fig. 8.1. However, it so happens that the human eye is extremely sensitive to just such changes of gradient, and therefore the *visual* appearance of a photographic plate is that of a number of broad, but well-defined, dark rings on a lighter background even though the *physical* gradation of blackening is as shown in Fig. 8.1. The radii of these visually observed diffraction rings are found to correspond with the maxima of the second term in 8.8′, i.e. they occur where

$$\sin \mathscr{S} l_{ij} = 1.$$

From 8.5′ we get

$$l_{ij} = (4n + 1)\lambda L/4R \qquad 8.9$$

where n is an integer.

Similarly, the radii of the apparently dark rings between adjacent bright rings correspond to

$$\sin \mathscr{S} l_{ij} = -1$$

giving

$$l_{ij} = (4n + 3)\lambda L/4R \qquad 8.9'$$

The different interatomic distances existing in the molecule can therefore be determined by simple visual measurement of the diffraction pattern. It is found that the apparent visual intensity of a diffraction ring corresponding to an interatomic distance l_{ij} is, very roughly, proportional to $Z_i Z_j$. This helps in assigning a measured l_{ij} to the appropriate pair of atoms in the molecule, and therefore in determining the precise arrangement of the atoms.

EXAMPLE

A diffraction pattern of $SiCl_4$ vapour made by electrons of wavelength $\lambda = 6{\cdot}06$ pm and a camera length L of 121·9 mm appears to consist of bright and dark rings having the following radii, measured in mm.*

Bright	2·8		**5·0**		7·4		9·5		**11·6**		14·0		**16·3**
Dark		3·9		6·3		8·5		10·6		12·9		—	
Bright			18·7										
Dark	17·5												

The rings whose radii appear in heavy type were particularly strong.

Show that these observations are consistent with a molecular structure of 4 chlorine atoms arranged tetrahedrally round a silicon atom, the Cl–Cl distance being about 328 pm and the Si–Cl distance about 201 pm.

Note: There are 6 Cl–Cl bonds and 4 Si–Cl bonds.

Since the atomic numbers of Si and Cl are 14 and 17 respectively, the intensity ratio of the Cl–Cl terms to the Si–Cl terms is

$$\frac{17 \times 17 \times 6}{14 \times 17 \times 4} \sim 2$$

Hence the Cl–Cl interference terms dominate the pattern. The effect of the Si–Cl interference terms is to give an extra intensity to certain rings where there is an approximate coincidence of the Cl–Cl and Si–Cl maxim.

* These data have been computed from results published by Brockway and Wall [8].

8.5 *Interpretation of diffraction patterns of molecular gases: radial distribution method*

The early work on the determination of molecular structure was all done by visual measurements of the photographic plate, the observations being interpreted as described in § 8.4. Later workers have sought a greater precision in their determinations of interatomic spacings by the use of a microphotometer to measure their plates.

The usual technique is to mount the plate in a special holder on the microphotometer so that during the measurement it can be

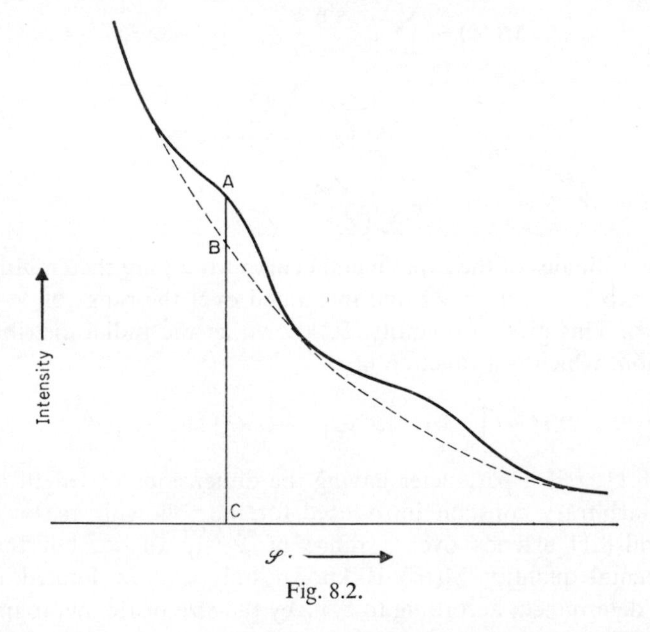

Fig. 8.2.

rapidly spun about an axis coinciding with the centre of the pattern. The microphotometer record obtained in this way represents the radial variation of blackening of the photographic plate averaged over all azimuths. Combining the data of this record with the exposure-blackening characteristic of the photographic emulsion, and introducing the scale factor $2\pi/\lambda L$ (equation 8.5′) to convert distances on the photographic plate to $\mathscr{S}$ values, one obtains a curve as shown schematically in Fig. 8.2 relating the scattered intensity to $\mathscr{S}$. As pointed out in § 8.4, the effect of the molecular interference term is simply to produce a small inflection in a monotonic curve. One draws in the background curve as shown dotted in Fig. 8.2 and then, corresponding to any $\mathscr{S}$ value, calculates the quantity

$$\mathrm{M}(\mathscr{S}) = \frac{\mathrm{AC}}{\mathrm{BC}} - 1 = \frac{\mathrm{AB}}{\mathrm{BC}}.$$

It is easily seen that $\mathrm{M}(\mathscr{S})$ is the ratio of the molecular interference term in equation 8.8 to the other terms. Hence the experimentally determined quantity $\mathrm{M}(\mathscr{S})$ can be expressed as

$$M(\mathscr{S}) = \sum_{ij} c_{ij} \frac{\sin \mathscr{S} l_{ij}}{\mathscr{S} l_{ij}} \exp \{-\mathscr{S}^2 \Delta_{ij}{}^2/4\}, \qquad 8.10$$

where
$$c_i = \frac{(Z_i - F_i)(Z_j - F_j)}{\Sigma\, [Z_i - F_i)^2 + Z_i q_i]}.$$

For large $\mathscr{S}$
$$c_{ij} \rightarrow \frac{Z_i Z_j}{\Sigma\, (Z_i{}^2 + Z_i)}.$$

The ordinates of the experimental curve $\mathrm{M}(\mathscr{S})$ are then multiplied by $\mathscr{S} \exp \{-b\mathscr{S}^2) \sin \mathscr{S} r$ and integrated over the range of $\mathscr{S}$ from 0 to ∞. This gives a quantity D, known as the radial distribution function, which is a function of r

$$D(r) = \int_0^\infty M(\mathscr{S})\mathscr{S} \exp \{-b\mathscr{S}^2\} \sin \mathscr{S} r \, d\mathscr{S}. \qquad 8.11$$

In 8.11, r is a parameter having the dimensions of length and b is an arbitrary constant introduced for the following reason. The integral 8.11 extends over a range of $\mathscr{S}$ up to ∞, but the experimental quantity $\mathrm{M}(\mathscr{S})$ is known only over a limited range of $\mathscr{S}$ determined, according to 8.5′, by the size of the photographic plate. The factor $\exp \{-b\mathscr{S}^2\}$ is therefore introduced into 8.11, the constant b being so chosen that there is only a negligible contribution to the integral from $\mathscr{S}$ values outside the experimental range.

Figure 8.3 shows schematically the form of the radial distribution curve corresponding to a measured scattering curve such as Fig. 8.2. In order to get a simple picture of the significance of the radial distribution function, we will consider the contribution to this function from a single interatomic interference, i.e. a single term ij

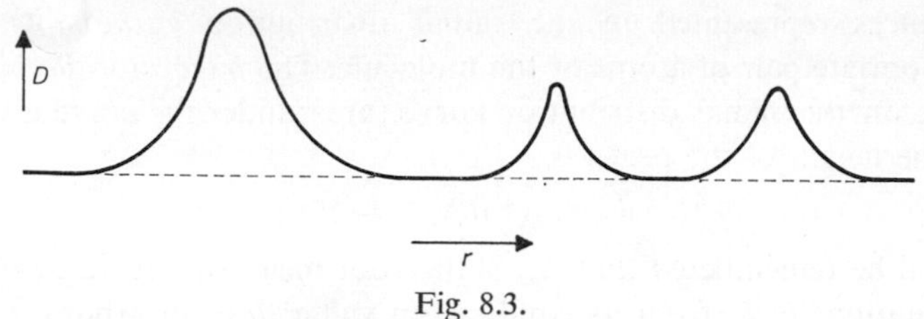

Fig. 8.3.

of 8.10. If we insert in 8.11 a typical term from 8.10 we have, dropping the suffices i, j.

$$D(r) = \int_0^\infty c \frac{\sin \mathscr{S} l}{l} \exp\{-\mathscr{S}^2(\Delta^2/4 + b)\} \sin \mathscr{S} r \, d\mathscr{S}.$$

Over the greater part of the pattern, c is approximately constant. Hence

$$D(r) = \frac{c}{2l}\int_0^\infty \{\cos \mathscr{S}(r - l) - \cos \mathscr{S}(r + l)\} \exp\{-\mathscr{S}^2(\Delta^2/4 + b)\} \, d\mathscr{S}.$$

Expressing the cosine terms in exponential form using Euler's identity (cf. equation 8.7) gives

$$D(r) = \frac{c\sqrt{\pi}}{2l(\Delta^2 + 4b)^{\frac{1}{2}}} [\exp\{-(r - l)^2/(\Delta^2 + 4b)\} - \exp\{-(r + l)^2/(\Delta^2 + 4b)\}]. \quad 8.12$$

If we restrict ourselves to the range of positive values of r, the second term is negligible.

We have here studied the contribution of a *single term* of 8.10 to $D(r)$. It is therefore clear that the complete radial distribution function is

$$D(r) = \sum_{i,j} \frac{c_{ij}\sqrt{\pi} \exp\{-(r - l)^2/(\Delta_{ij}^2 + 4b)\}}{2(\Delta_{ij}^2 + 4b)^{\frac{1}{2}} l_{ij}}. \quad 8.13$$

The radial distribution function therefore consists of a number of bell-shaped, Gaussian curves having maxima at values of r equal to each interatomic distance l_{ij}. The area under each such curve is

$$\pi c_{ij}/2l_{ij}$$

and is therefore determined by the atomic numbers of the corresponding atoms: this helps to correlate the various interatomic distances represented in the radial distribution curve with the appropriate pair of atoms of the molecule. The *integral width* of any peak on the radial distribution curve (area under the curve divided by the height of the peak) is

$$w_{ij} = \sqrt{\pi}(\Delta_{ij}^2 + 4b)^{\frac{1}{2}}. \qquad 8.14$$

It will be remembered that Δ_{ij} is the root mean square value of the fluctuations of l_{ij} from its equilibrium value; b is an arbitrary constant introduced in the computation of $D(r)$ from the experimental $M(\mathscr{S})$ (equation 8.11) to reduce to negligible amount the contributions from the range of $\mathscr{S}$ values beyond those experimentally accessible. Usually, the two terms in 8.14 are of comparable magnitude. The magnitude of the molecular vibrations Δ_{ij} can be calculated from the integral width of the peak on the radial distribution curve using the relation

$$\Delta_{ij} = (w_{ij}^2/\pi - 4b)^{\frac{1}{2}}. \qquad 8.15$$

It is now possible, using the radial distribution curve method, to determine interatomic distances in gas molecules to an accuracy of about 0·1%, and to provide useful information about the form of the molecular vibrations.

8.6 *Refinements of the radial distribution curve method*

The treatment of the previous section is based on the early work of Debye [17, 18]. Refinements have been introduced by, among others, Bartell, Brockway and Schwedeman [4] and Karle and Hauptman [34]. The former authors take account of the fact that the quantities c_{ij} of 8.10 are not constant at low $\mathscr{S}$ values. They correct the experimental $M(\mathscr{S})$ function to a function $M_c(\mathscr{S})$ which is given by equation 8.10 with the c_{ij} replaced by constants equal to their limiting values. The correction term is small, and can be calculated with sufficient accuracy when an approximate structure of the molecule is known; in extreme cases, a method of successive approximations can be used.

Bartell [3] has given a critical discussion of the precise significance of the interatomic distances l_{ij} given by the radial distribution method in terms of the potential energy–distance curve of the two atoms involved.

9

Electron Interference

9.1 *Introduction*

The wave nature of electrons is well demonstrated by the *diffraction* of electrons by regular lattices of atoms in crystals, a subject which is the main theme of this book. Nevertheless, it is obviously of scientific interest to consider whether these wave properties can be demonstrated by other experiments more closely analogous to the optical experiments which provided the experimental foundation of

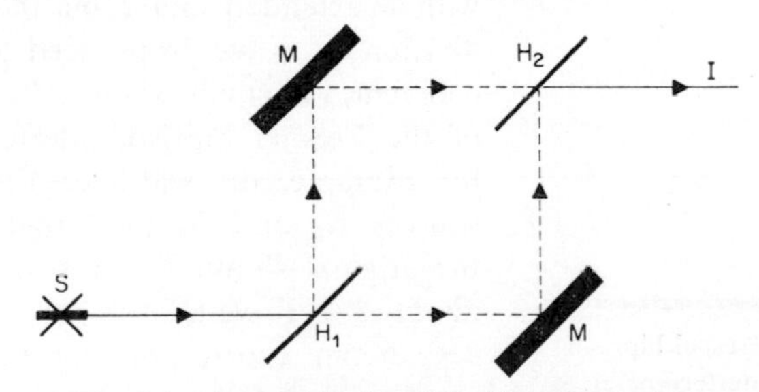

Fig. 9.1. Mach–Zehnder optical interferometer.

S Source of light; H_1, H_2 Half silvered mirrors; M Full silvered mirrors. The incident beam of light from the source S is shown by a full line. The half silvered mirror H_1 divides this into two separate beams which are shown as dashed lines. These two beams are reunited by the half silvered mirror H_2 to form the emerging beam I. Interference fringes are observed when the emerging beam enters a telescope.

the wave theory of light. In particular, it would be of interest to observe *interference* phenomena arising from the superposition of only a few waves.

It is customary to classify optical interferometers according to whether they produce the interfering beams by *division of amplitude* or *division of wave front*. Typical of the first type are the Mach–Zehnder and Michelson interferometers, of which the former is illustrated in Fig. 9.1. Typical of the second type is the Fresnel biprism interferometer (Fig. 9.2). Each half of the biprism deviates all the rays passing through it by the same amount, and as a result produces a virtual image S_1 or S_2 of the source S. The waves from the two virtual sources S_1 and S_2 produce interference fringes in any plane. It is shown in textbooks of optics that the separation Δ of the fringes in any plane P is given by

$$\Delta = \lambda L/D, \qquad 9.1$$

where D is the distance S_1S_2 between the virtual sources and L is the distance of these sources from the plane P in which the fringes are formed.

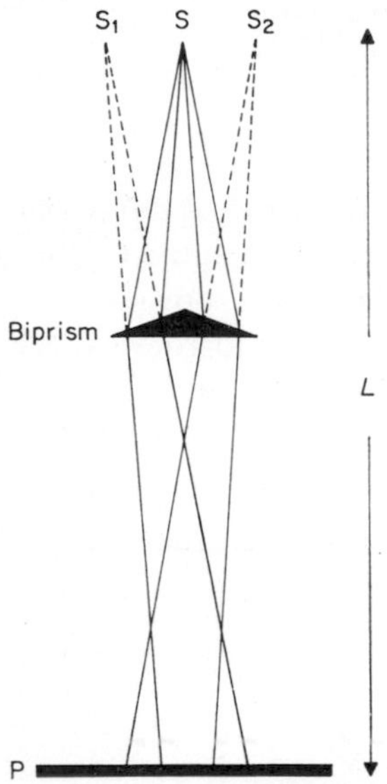

It is well known that interferometers using division of amplitude may be used with an extended source, but those using division of wave front need a highly coherent, i.e. small, source. In the case of the Fresnel biprism interferometer, for example, the width of the virtual sources S_1 and S_2 (and therefore of the original source S) must be less than about $\Delta/4$. If we take as typical values $\lambda = 6$ pm (corresponding to 40 kV electrons) and $L = 500$ mm, then if $D \sim 8\ \mu$m, the width of the source must be less than about 100 nm. The difficulty of obtaining sufficient intensity from such a narrow source naturally led to the first successful attempt at constructing an electron interferometer being based on the principle of division of amplitude rather than division of wave front.

Fig. 9.2. Fresnel biprism optical interferometer.

9.2 *Division of amplitude: the three-crystal interferometer*

The principle of the three-crystal interferometer (Marton *et al.* [44]) is shown in Fig. 9.3. Three thin single-crystal films M_1, M_2, M_3 are used. An electron beam incident on such a film is partly transmitted without deviation and partly diffracted: the film therefore acts as the analogue of a half silvered mirror. It is obvious from a comparison of Figs. 9.3 and 9.1 that the device is, in principle, the electron analogue of a highly skewed Mach–Zehnder interferometer.

The crystals were of copper grown epitaxially on rock salt, which was subsequently dissolved. These single crystal films were about 20 nm thick. They could either be mounted on a standard electron microscope grid 3 mm in diameter, or they could be mounted across a 1-mm hole in a metal foil. The fringes had a spacing of 165 nm and were observed with an electron microscope (the interferometer being placed between the condenser and objective).

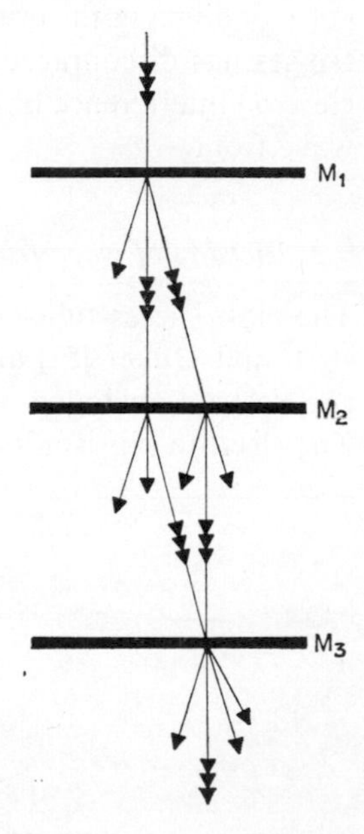

Fig. 9.3. The three-crystal electron interferometer. The three crystal films M_1, M_2 and M_3 each diffract the beams as shown (for simplicity, only the zero- and first-order beams are shown). The lines marked with triple arrows show how the incident beam is divided into two beam which are recombined at M_3.

The separation of the crystal films was about 35 mm.

The most significant outcome of experiments with this interferometer was the observation that interference fringes could still be observed when the path difference between the two interfering beams was about 6000 wavelengths, or 28 nm. This sets a lower limit to the length of the electron wave trains, and was the first direct experimental determination of this quantity.

Theoretically, the length of the wave train is expected to be much greater than this. A Fourier analysis of a harmonic wave train containing N waves of wavelength λ shows that the bulk of the energy is contained in a wavelength range $\lambda \pm \lambda/N$. From equation 1.3, we see that such a range of wavelengths is produced by a fluctuation of the accelerating potential P of $\pm 2P/N$. Since the short period stability of a high voltage supply is typically one part in 10^4 to 10^5, or better, this would correspond to a wave train of 2×10^4 to 2×10^5 wavelength. It would therefore seem that it is imperfections of the interferometer which limit the path difference over which interference can be observed.

The three-crystal interferometer is of historical interest as being the first successful electron interferometer. However, the great difficulty of aligning the crystals and obtaining sufficient intensity in the fringes has discouraged further development, and later studies of electron interference have all made use of the principle of division of wave front.

9.3 *Division of wave front: the Möllenstedt biprism interferometer*

The most successful electron interferometer was devised by Möllenstedt and Düker [51] and further developed by Keller [36]. This is the electron analogue of the Fresnel biprism optical interferometer. The electron biprism itself (Fig. 9.4) consists of a quartz fibre about

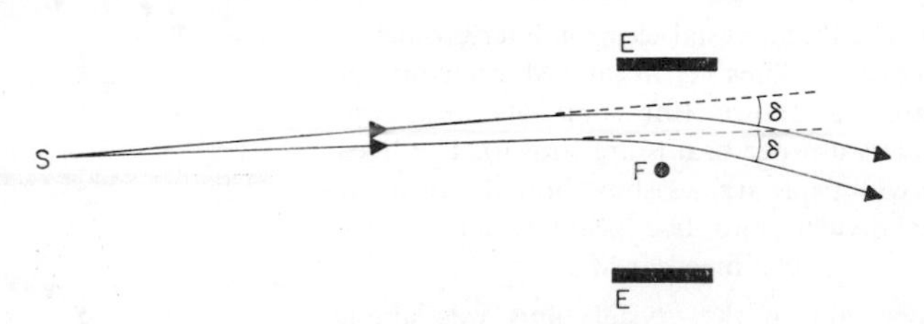

Fig. 9.4. Möllenstedt biprism.

F Gold-coated quartz fibre, 1 μm diameter (greatly exaggerated in figure); E Earthed metal plates.

The figure shows two electron trajectories from a source S. The deviation δ is constant irrespective of the closest distance of approach of the electron to the fibre.

1 μm diameter coated with gold by vacuum deposition so as to make it conducting. This is mounted midway between two earthed metal plates 5 mm long and 4 mm apart. If the potential of the fibre is a few volts positive to earth, electrons passing near it are deviated through an angle which is independent of the perpendicular distance of the electron trajectory from the fibre (see Fig. 9.4). The device therefore acts as the precise analogue of the optical biprism.

The principle of the complete instrument is shown in Fig. 9.5. The discussion of § 9.1 shows that for a satisfactory electron optical interferometer one needs a line source of electrons less than about 100 nm wide. This is obtained by imaging the cross-over of a conventional electron microscope gun A by means of two cylindrical

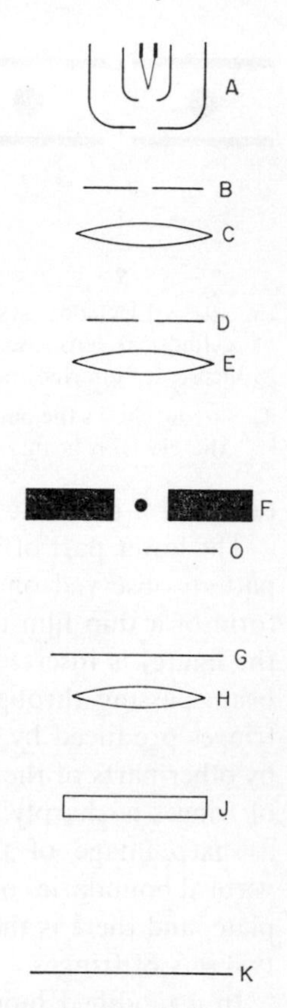

Fig. 9.5. Möllenstedt interferometer.

A Electron gun; B Slit 300 μm wide; C Cylindrical lens: $f = 3{\cdot}6$ mm; D Slit 80 μm wide; E Cylindrical lens: $f = 5{\cdot}2$ mm; F Biprism; G Removable fluorescent screen; H Cylindrical lens: $f = 3{\cdot}2$ mm; J Cylindrical lens with axis in the plane of the figure: $f = 5{\cdot}2$ mm $- \infty$; K Fluorescent screen or photographic plate for recording final image; O Object.

electrostatic lenses C and E, which form a line image of the cross-over about 50 nm wide just above the biprism F. The cylindrical lenses, illustrated in Fig. 9.6, each consist of two parallel metal rods 3 mm diameter and 16 mm apart mounted midway between two earthed metal plates. The rods are connected to an adjustable source of high potential.

If it is desired to place an object in one of the two interfering beams produced by the biprism, then this object is introduced at O. Below the object are placed two further electrostatic cylindrical lenses H and J. The lens J is mounted with its axis in the plane of the figure, i.e. at right angles to the axes of the other lenses and the fibre of the biprism. Its purpose is to image the object on to the photographic plate K in a direction perpendicular to the plane of the figure. The magnification in this direction is about $\times 33$. In the plane of the figure, the rays pass through J without deviation. Non-localized fringes in a plane above the lens H are imaged by this lens on to the fluorescent screen or photographic plate K with a magnification $\times 800$. In an alternative version of this interferometer, the positions of the lenses H and J are interchanged, but the general principle of the instrument is the same.

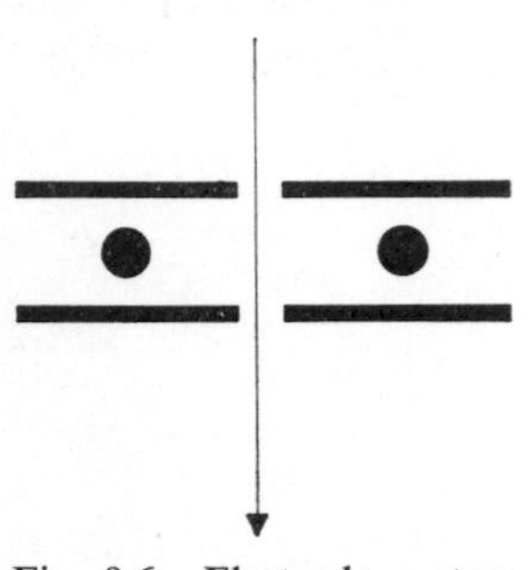

Fig. 9.6. Electrode system of cylindrical lens used in Möllenstedt interferometer. The arrow shows the path of the electron beam.

The lower part of Fig. 9.7(a) indicates the appearance of the fringe pattern observed on the photographic plate when an object in the form of a thin film of square shape (indicated in the upper part of the figure) is inserted in one of the interfering beams. The electron beam passing through the object is retarded in phase, causing the fringes produced by it to be displaced with respect to those formed by other parts of the same beam. The boundary between the two sets of fringes is sharply defined because the lens J of Fig. 9.5 produces a sharp image of the horizontal boundaries of the object. The vertical boundaries of the object are not imaged on the photographic plate, and there is therefore no sharp vertical boundary between the two sets of fringes.

In a modified biprism instrument devised by Buhl [9] the lenses H and J of Fig. 9.5 are replaced by a conventional magnetic electron microscope objective lens. This forms a sharp image of an object, so that an irregular object such as the one indicated in the upper part of Fig. 9.7(b) gives rise to the fringe system shown in the lower part of this figure.

The advantage of Keller's arrangement is that, by reducing the magnification in one direction, the brightness of the final pattern is greatly increased. Its disadvantage is that it can be used only to study 'artificial' objects such as thin films prepared by vacuum deposition on to a supporting substrate through a small square hole in a mask. Buhl's arrangement, on the other hand, removes this limitation on the geometry of the object at the expense of brightness of the pattern.

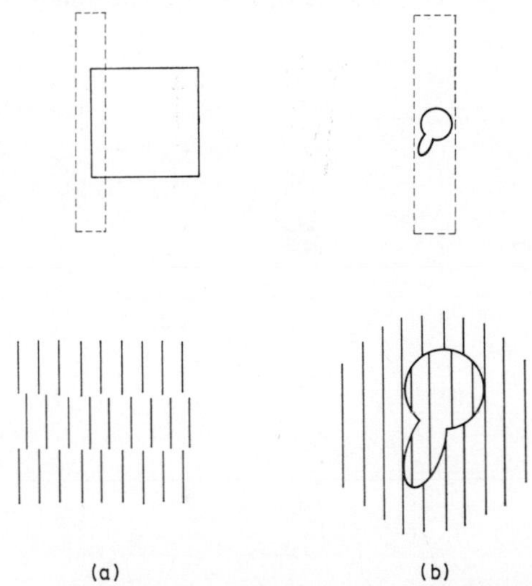

Fig. 9.7. General appearance of interference micrographs obtained with a Möllenstedt interferometer. The upper part of each figure shows the object, indicated by a solid outline, on which is superposed a broken outline representing the shadow of the biprism in the object plane.

(a) Interferometer using cylindrical lenses. The horizontal boundaries of the object are sharply defined in the image, being the junctions of the two sets of fringes. The vertical boundaries of the object are not represented in the image.
(b) Interferometer using spherical lenses. An irregular object can now be sharply imaged.

9.4 *Applications of the biprism interferometer*

The first use made of the biprism interferometer by Möllenstedt and his co-workers was to determine, to an accuracy of $\pm\frac{1}{2}\%$, the wavelength of the electrons without recourse to a diffracting crystal. This determination was based on equation 9.1 and requires, in addition to measurements of the fringe separation and the distance from the virtual sources to the plane in which the interference fringes are observed, a knowledge of the distance apart of the virtual sources formed by the biprism. Figure 9.8 shows how this is done. The edges AA′ of the shadow of the biprism fibre cast by the source S are noted when the fibre is at zero potential. The potential of the fibre is then raised to its normal operating value, causing the source S to be replaced by the virtual sources S_1 and S_2. The shadow edges B and

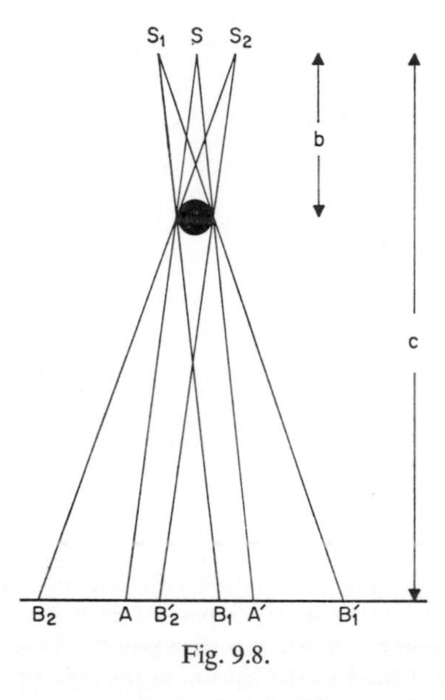

Fig. 9.8.

B′ cast by these sources are observed. Then, from the geometry of the figure, it follows that if

$$d = AB_1 = A'B_1' = AB_2 = A'B_2'$$

the separation D between the virtual sources S_1 and S_2 is given by

$$D = \frac{2bd}{c - b}. \qquad 9.2$$

The distance b between the source S (which, it will be recalled, is an electron optical image of the cross-over of the electron gun formed by two electron lenses) and the biprism is, from the geometry of the figure, given by

$$b/c = 2r/\delta, \qquad 9.3$$

where δ is the width AA′ of the shadow of the earthed biprism fibre cast by the source and r is the radius of the biprism fibre. From 9.1, 9.2 and 9.3 we see that the electron wavelength can be determined entirely from measurements made with a travelling microscope.

The next application of the interferometer was to measure the coherence length of the electron wave train. For this purpose, a small adjustable potential difference was applied between the side plates of the biprism, thus moving the whole fringe system sideways. In this way, Keller [36] showed that a total of 2200 interference fringes existed in the pattern, corresponding to a wave train length of 2200λ or 14·2 nm for the 35-kV electron beam used.

The most extensive measurements made with the interferometer have been those of the inner potentials of various materials. The principle of such measurements has been hinted at in the previous section, in which the instrument was described. A thin film of the material, having a sharp boundary between two parts of different thickness, is mounted below the biprism. From a pattern such as that illustrated schematically in Fig. 9.7(a) the displacement n of one set of fringes with respect to the other is measured in terms of the fringe spacing as unit. If t is the thickness of the film and μ its refractive index for electrons

$$n = (\mu - 1)t/\lambda. \qquad 9.4$$

The inner potential is then given in terms of the refractive index from equation 5.3. The determination of n obviously requires that the zero-order fringe should be identified in each system of fringes. This is done by superposing a small alternating potential on the steady potential applied to the biprism fibre. The zero-order fringe remains sharply defined, but the remaining fringes become blurred. The reader will doubtless observe a certain similarity between this technique and the use of white light to locate the zero-order optical interference fringe when using a Michelson interferometer.

The thickness of the film is estimated by depositing under as nearly as possible the same conditions a similar film on a glass substrate. Silver is then vacuum deposited over the film and the height of the step occurring in the silver surface at the edge of the film is measured by optical interferometry. This method of measuring inner potentials is capable only of poor accuracy – about $\pm 15\%$ – mainly because of the difficulty of measuring the film thickness.

9.5 *Effect of magnetic vector potential on the phase of electron waves*

The biprism interferometer has been used to verify a rather surprising prediction of Aharonov and Bohm [1] that a magnetic field can influence electron waves even when these waves are confined to regions of space into which the magnetic field does not extend. To understand this theory, let us consider the arrangement in Fig. 9.9. A magnetic field of flux density B, perpendicular to the plane of the figure, exists in the small cylindrical region indicated; the field elsewhere is supposed to be negligible. Such a limited field can easily be produced by a long solenoid carrying a current. A biprism interferometer is set up so that electrons from a source S can reach a plane I, where interference fringes can be observed, by two routes a and b, neither of which is exposed to the magnetic field.

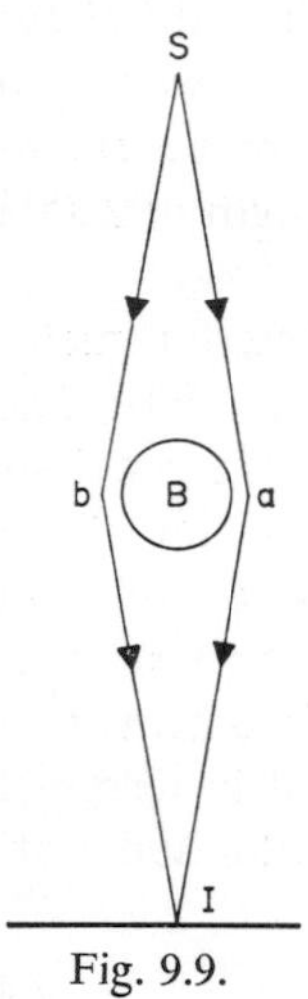

Fig. 9.9.

The Schrödinger equation for an electron in a field-free space can be written

$$-\hbar^2\nabla^2\psi = 2mE\psi, \qquad 9.5$$

where E is the energy of the electron.

A magnetic field $\mathbf{B}$ can be defined by a vector potential $\mathbf{A}$:

$$\mathbf{B} = \text{curl}\,\mathbf{A} \qquad \text{div}\,\mathbf{A} = 0. \qquad 9.6$$

In the presence of such a magnetic field, the Schrödinger equation 9.5 becomes

$$-\hbar^2\nabla^2\psi + 2\hbar ie\mathbf{A}\cdot\text{grad}\,\psi + e^2\mathbf{A}^2\psi = 2mE\psi. \qquad 9.7$$

In any region where there is zero magnetic field (and this applies to all the space accessible to electrons travelling by routes a or b in Fig. 9.9) we have curl $\mathbf{A} = 0$. We can therefore express $\mathbf{A}$ as the gradient of a scalar function α:

$$\mathbf{A} = \text{grad}\,\alpha. \qquad 9.8$$

It can then be shown that if ψ_0 is a solution of 9.5, the solution of 9.7 is

$$\psi_0 \exp -(ie/\hbar)\alpha. \qquad 9.9$$

The phase of the electron wave at I which has travelled by route a

is therefore changed with respect to the wave at S by

$$\delta_a = (e/\hbar)(\alpha_I - \alpha_S)$$

where α_I and α_S are the values of α at I and S respectively.

Using 9.8, we can write this phase difference as

$$\delta_a = (e/\hbar)\int_a \mathbf{A}\cdot d\mathbf{l} \qquad 9.10a$$

where the line integral of **A** is taken along a route such as *a*.

In the same way, the phase of the wave travelling by route *b* is changed with respect to the wave at S by

$$\delta_b = (e/\hbar)\int_b \mathbf{A}\cdot d\mathbf{l}. \qquad 9.10b$$

Hence the phase difference between two waves reaching I by the two routes *a* and *b* is

$$\delta_a - \delta_b = (e/\hbar)\oint \mathbf{A}\cdot d\mathbf{l},$$

where the line integral is taken round the circuit S*a*I*b*S. Using 9.6 we can express this phase difference as

$$\delta_a - \delta_b = (e/\hbar)\int \mathbf{B}\cdot d\mathbf{S}$$

$$= (e/\hbar)\,\Phi, \qquad 9.11$$

where $d\mathbf{S}$ is an element of area of the circuit S*a*I*b*S and $\oint = \int \mathbf{B}\cdot d\mathbf{S}$ is the total magnetic flux through this circuit, i.e. the magnetic flux in the cylindrical region indicated in Fig. 9.9.

Möllenstedt and Bayh [50] were able to confirm this prediction in a very convincing manner. They used a biprism interferometer in which the maximum separation of the interfering beams was 60 μm. In the space between the beams they placed a miniature solenoid only 15 μm diameter. On increasing the current through the solenoid slowly from zero, they observed a progressive shift of the interference fringes. According to 9.11, a shift of one fringe is produced by a magnetic flux of h/e. Hence by measuring the current in the solenoid required to produce a shift of one fringe, it is possible to determine the quantity h/e. The value found by Möllenstedt and Bayh was within 14% of the true value and was regarded as a satisfactory confirmation of the theory of Aharonov and Bohm.

10

Electron Diffraction Effects in the Electron Microscope

10.1 *Introduction*

The smallest structures studied in the electron microscope are roughly 100 times larger than the wavelength of the electrons. Diffraction effects will therefore result in a scattering of a parallel beam of electrons through angles of the order of 10^{-2} radian. Since the objective aperture of an electron microscope accepts only electrons contained in a cone of semi-angle $\sim 10^{-3}$ radian, it follows that many of the electrons diffracted by the object will fail to enter the objective aperture and will therefore not contribute to the intensity of the image. Indeed, the contrast, i.e. the variation of intensity from one point to another, of an electron microscope image is mainly due not to variations in opacity of different parts of the specimen but rather to their differing power of scattering or diffracting electrons. It follows that some of the details of electron microscope pictures require for their interpretation a knowledge of the theory of electron diffraction. In this chapter, we shall briefly survey some of the effects observed. For a more detailed discussion of these topics, the reader is referred to the book by Hirsch *et al.* [31].

10.2 *Fresnel fringes*

The simplest diffraction effect is the Fresnel fringes which can be observed when the specimen consists of a fairly opaque film with holes in it. Thick carbon films, of the kind commonly used as specimen substrates, frequently have a few small holes which serve admirably. If the image of a hole is observed when the microscope is slightly out of focus, a few bright fringes will be seen running parallel to the edge of the hole. These fringes are produced by diffraction of the electrons at the edge of the hole. The theory of the diffraction of waves at a straight-edge is given in textbooks on optics (e.g. Ditchburn [19]) and only the essential results will be quoted here. Let a plane wave be propagated past a straight-edge S as indicated in Fig. 10.1. We consider the intensity at a point P in a plane OP

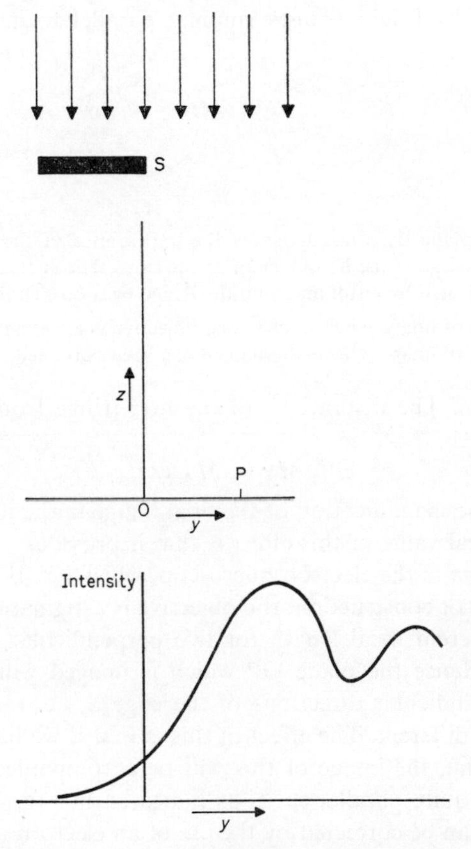

Fig. 10.1. Fresnel diffraction at a straight-edge.

distance z from the edge. O is the position of the shadow of the edge in the absence of diffraction, and we denote by y the distance OP. Then the theory of diffraction asserts that the variation of intensity in the plane OP is as indicated at the bottom of the figure. The position P of the first intensity maximum is such that $SP - SO \approx \lambda/2$.

If OP is small compared with SO, this condition becomes

$$y \approx \sqrt{(\lambda z)}. \qquad 10.1$$

Applying this result to the electron microscope, we see that if the microscope is not focussed exactly on the plane of the specimen S, but on a plane OP, then the (slightly blurred) image of the edge is accompanied by bright fringes running parallel to it, as indicated

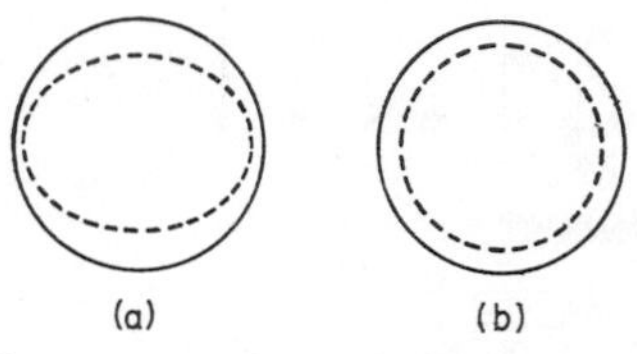

Fig. 10.2. Schematic diagrams to show the appearance of the Fresnel bright fringe near the image of the boundary of an aperture. The aperture boundary is indicated by a full line, and the fringe by a broken line.

(a) Appearance of image when microscope objective is astigmatic.
(b) Appearance of image when astigmatism has been corrected.

in Fig. 10.2(b). The distance Y of the first fringe from the edge is given by

$$Y = My = M\sqrt{(\lambda z)} \qquad 10.2$$

where M is the magnification of the final image in the microscope.

The practical value of this effect is that it provides a sensitive test for *astigmatism* in the electron microscope objective. If, due to small imperfections of construction, the objective is astigmatic, it will have a slightly different focal length for two perpendicular directions of the fringes. Hence the plane OP which is imaged will be different for two perpendicular directions of the edge S, i.e. the values of z and y will be different. The effect of this is that if we have a circular hole in the film, the image of this will be accompanied by a fringe which is not quite parallel to it, as indicated in Fig. 10.2(a). Such astigmatism can be corrected by the use of an electron optical device called a stigmator placed immediately after the objective. This device can be regarded as equivalent to a pair of crossed cylindrical lenses of equal and opposite power. The power of the cylindrical lenses, and the orientation of their axes, can be adjusted at will. With this device, any astigmatism of the objective can be counteracted. When this has been done, the Fresnel fringe runs exactly parallel to the image of the hole in the specimen film (Fig. 10.2(b)).

10.3 *Thickness and bend contour fringes*

It has been pointed out in § 10.1 that contrast in an electron microscope image (particularly of a crystalline specimen) arises because any electrons diffracted by the crystal are unable to enter the objective aperture. The intensity at any point in the image therefore depends on the intensity of the direct transmitted beam at the lower surface of the crystal. This is readily calculated in the case of a crystalline lamina. Equation 6.41 gives the intensity of the diffracted beam on the two-beam approximation of the dynamical theory, i.e. on the assumption that the only beams of appreciable intensity are the direct beam and one diffracted beam. For a lamina set nearly normal to the incident beam ($\alpha = 0$), equation 6.41 becomes

$$|T|^2 = 1 - \frac{\sin^2 \pi SD}{S^2\Delta^2}, \qquad 10.3$$

where
$$S^2 = \frac{1}{\Delta^2} + (\varepsilon g)^2. \qquad 6.37$$

If the crystal is not of uniform thickness, the transmitted intensity will be different at different parts. For a wedge-shaped crystal, we shall observe parallel fringes of constant thickness. Such fringes of constant thickness are often seen in electron micrographs of thin crystalline films. Equations 10.3 and 6.37 show that if the crystal is oriented at exactly the Bragg angle ($\varepsilon = 0$), the thickness difference between adjacent fringes is equal to the extinction distance Δ, typically a few tens of nm. Near the edge of a film, the opposite faces of the film are typically inclined to each other at an angle of a few degrees. This means that the distance apart of the fringes is of the order of $\frac{1}{2}$ μm, so that they are easily visible under fairly low magnification.

Equation 10.3 further shows that if the film is of constant thickness, the intensity of a transmitted beam depends on S and therefore on ε. If, therefore, the film is slightly curved, we shall observe fringes of constant inclination, or *bend contours*. For larger values of ε, $S \approx \varepsilon g$, and the second term of 10.3 approximates to the form

$$\left(\frac{\sin a\varepsilon}{a\varepsilon}\right)^2.$$

This is the same as the expression, given in any textbook of optics, for the intensity variation in the Fraunhofer diffraction pattern of a single slit. Qualitatively, therefore, the appearance of a bend contour is that of a central dark line (corresponding to $\varepsilon = 0$) fringed by less intense bright and dark lines.

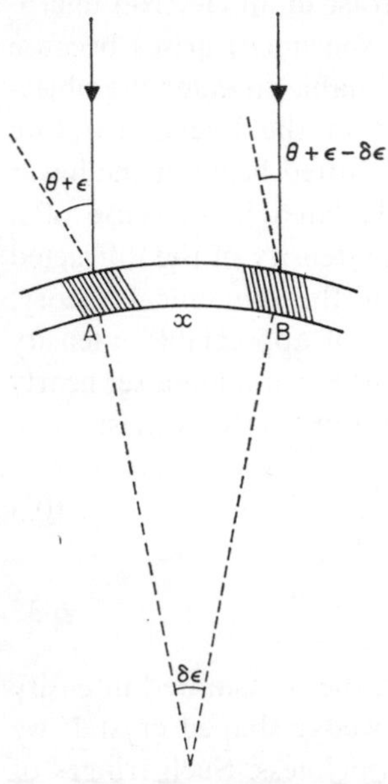

Fig. 10.3. The origin of the bright lines in bend contours. The transmitted beam has a high intensity at A and B, giving bright lines at these points.

Equation 10.3 shows that adjacent lines in a bend contour correspond to values of S differing by $1/D$. Making the approximation $S = \varepsilon g$, we find that adjacent lines have values of ε differing by

$$\delta\varepsilon = 1/gD = d/D,$$

where d is the interplanar spacing of the diffraction. Let us suppose the plane of the crystalline film is curved about an axis parallel to the reflecting planes. Consideration of Fig. 10.3 shows that if the distance apart of adjacent lines in the bend contour is x, and r is the radius of curvature of the film

$$x/r = \delta\varepsilon = d/D$$

so that $x = r\,d/D$. Typically, $d \sim 200$ pm, $D \sim 200$ nm, $r \sim 10$ mm, so that $x \sim 1$ μm.

10.4 *Contrast effects in dislocated crystals*

The lattices of many real crystals are usually not geometrically perfect, as we have hitherto assumed, but contain imperfections known as *dislocations*. The study of dislocations lies outside the scope of this book. For our present purposes, we can regard a crystal with a dislocation as divided by a plane into two almost perfect crystals. The lattice of the crystal on one side of the plane is displaced with respect to that of the crystal on the other side of the plane; the displacement is measured by a vector **R** known as the Burgers' vector. The fault plane and the Burgers' vector both have a simple relation to the crystal lattice. For example, in the case of face-centred cubic crystals, the fault plane is usually a (111) plane and the Burgers' vector has components referred to the crystal axes, $\pm(a/3)(1, 1, 1)$.

In electron micrographs of thin crystal films, contrast effects are observed at a fault plane, such as AB (Fig. 10.4) running from the top to the bottom surface of the film. We commence our discussion of the contrast effects by observing that the effect of moving a

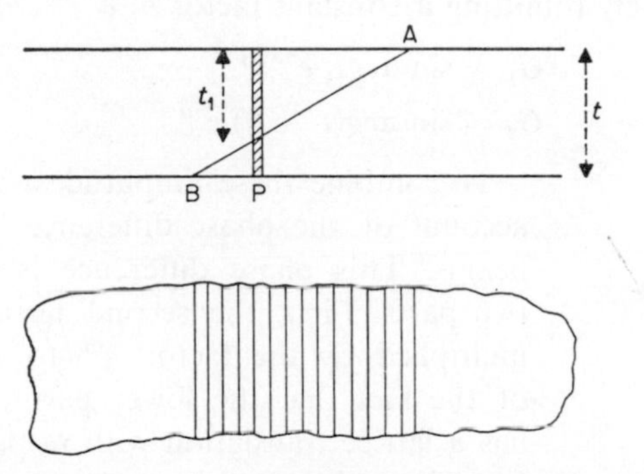

Fig. 10.4. The upper portion of the figure shows a crystal lamina of thickness t with a fault plane AB running through it. The lower figure is a schematic sketch of the electron micrograph of the crystal, showing fringes at the position of the fault plane.

crystal lattice through a vector distance **R** is to introduce a phase factor

$$e^{i\beta} = e^{2\pi i \mathbf{g} \cdot \mathbf{R}}$$

into the expression for the amplitude of the diffracted beam. This follows as an obvious generalization of the reasoning leading to equations 3.5 and 3.7. To determine the intensity of the beam which emerges from a point such as P on the lower surface of the film, we make the so-called column approximation. Since both the incident and diffracted beams make only a small angle with the normal to the film, any beam emerging at P must have passed through the portion of the crystal shown shaded. For a film of thickness ~100 nm, the width of this column of crystal is of the order of 2 nm.

The amplitude of the diffracted beam merging from P is made up of two terms: that due to diffraction in the portion, of thickness t_1, of the column lying above the fault plane AB; and that due to diffraction in the lower part of the column, of thickness $t - t_1$, where t is the thickness of the film. According to the simple kinematic theory, the amplitudes of these two terms can (apart from constant proportionality factors) be derived from equation 3.28a, which gives the amplitude of a diffracted beam emerging from a crystal lamina. In this equation, we write $\sec \alpha = 1$, since the electron beams are substantially normal to the film. The amplitudes diffracted by the upper and lower parts of the column in Fig. 10.4 are respectively (omitting a constant factor $b_1\, e^{-\pi i g}/\pi \varepsilon g$)

$$G_1 = \sin \pi\varepsilon g t_1\, e^{\pi i \varepsilon g t_1} \qquad 10.4$$

and

$$G_2 = \sin \pi\varepsilon g(t - t_1)\, e^{\pi i \varepsilon g(t-t_1)}$$

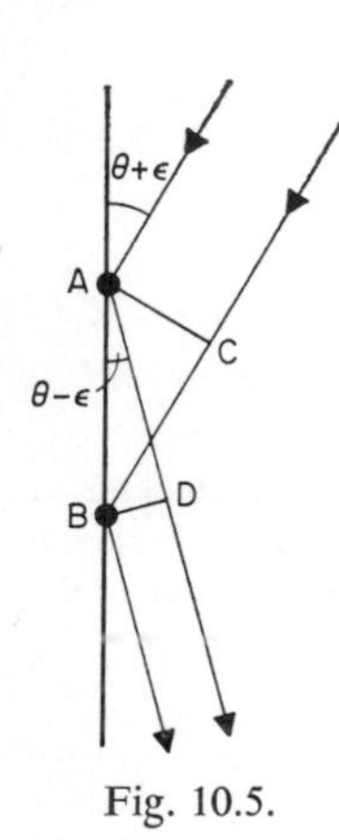

Fig. 10.5.

To combine these amplitudes, we must take account of the phase difference between these beams. This phase difference is made up of two parts. First, the second term G_2 must be multiplied by the factor $e^{i\beta}$ to take account of the fact that the lower part of the crystal has a lattice translation with respect to the top part. Secondly, the lower part of the crystal lies at a depth t_1 below the entrance surface. To evaluate this factor, consider two points A, B on a reflecting plane of atoms in the crystal (see Fig. 10.5). Let the incident ray have a glancing angle of $\theta + \varepsilon$, where θ is the Bragg angle. The diffracted ray will have a glancing angle $\theta - \varepsilon$. Then the path difference between the rays scattered at A and B is

$$\begin{aligned} AD - CB &= AB[\cos(\theta - \varepsilon) - \cos(\theta + \varepsilon)] \\ &= 2AB \sin\theta \sin\varepsilon \end{aligned}$$

The phase difference is therefore

$$\frac{2\pi}{\lambda} AB.2 \sin\theta \sin\varepsilon.$$

Using the Bragg relation, and writing $g = 1/d$ and $\sin \varepsilon \approx \varepsilon$, the phase difference is seen to be $2\pi AB g\varepsilon$. Since the reflecting planes are almost normal to the film, $AB = t_1$, and the phase factor with which G_2 must be multiplied to take account of the different depths of the two parts of the column is $e^{2\pi i g \varepsilon t_1}$. The resultant amplitude of the diffracted beam emerging from the point P in Fig. 10.4 is therefore

$$G_1 + e^{i\beta} e^{2\pi i g \varepsilon t_1} G_2 = e^{\pi i g \varepsilon t_1}[\sin \pi g \varepsilon t_1 + e^{i\beta} e^{\pi i g \varepsilon t} \sin \pi g \varepsilon (t - t_1)].$$

This corresponds to an intensity of the diffracted beam proportional to

$$G^2 = \sin^2 \pi g \varepsilon t_1 + \sin^2 \pi g \varepsilon (t - t_1) + \\ + 2 \sin \pi g \varepsilon t_1 \sin \pi g \varepsilon (t - t_1) \cos (\pi g \varepsilon t + \beta).$$

With a little trigonometrical manipulation, this can be transformed to

$$G^2 = \sin^2 (\pi g \varepsilon t + \beta/2) + \sin^2 \beta/2 + \\ + 2 \sin \beta/2 \sin (\pi g \varepsilon t + \beta/2) \cos 2\pi g \varepsilon Z, \qquad 10.5$$

where $Z = \frac{1}{2}t - t_1$ is the distance of the fault plane from the centre of the film. We observe that the intensity of the diffracted beam is a periodic function of the depth of the fault plane. The intensity of the direct, undiffracted beam, which alone enters the microscope objective, must exhibit a complementary periodic intensity. Hence an inclined fault plane such as is shown in Fig. 10.4 appears in the electron microscope as a set of cosine fringes.

10.5 *The imaging of atomic planes in the electron microscope*

In discussing the resolving power of an electron microscope, one usually starts by observing that a point source of electrons in the object is reproduced as a blurred *disc* of diameter D in the final image. If M is the overall magnification of the microscope, the resolution limit is

$$\delta = D/M.$$

This blurring of the image of a point object is due to two principal causes: (*a*) diffraction at the aperture of the objective lens, and (*b*) spherical aberration of the objective lens. The effect of diffraction by the objective aperture is diminished when the diameter of this aperture is increased. On the other hand, the effect of spherical aberration is more pronounced when the aperture is enlarged. It follows that there is an optimum size of aperture for minimum resolution limit. It is shown in textbooks on electron microscopy (e.g. Hirsch *et al.* [31]) that this minimum resolution limit is

$$\delta_{\min} \approx \lambda^{\frac{3}{4}} C_s^{\frac{1}{4}}, \qquad 10.6$$

where C_s is the spherical aberration constant of the objective. The corresponding optimum aperture is one which accepts from a point of the object a cone of rays of semi-angle β where

$$\beta \approx \lambda^{\frac{1}{4}} C_s^{-\frac{1}{4}}. \qquad 10.7$$

Typically, for 100 kV electrons,

$$\begin{aligned} C_s &= 3{\cdot}3 \text{ mm} \\ \lambda &= 3{\cdot}7 \text{ pm} \\ \delta_{\text{min}} &\sim 600 \text{ pm} \\ \beta &\sim 6 \times 10^{-3} \text{ rad.} \end{aligned} \qquad 10.8$$

δ_{min} is thus of the order of magnitude of the spacing of the planes of atoms in a crystal, so that serious consideration must be given to the possibility of 'seeing' the lattice of a crystal in the electron microscope.

The first point to be noted is that equation 10.6 will *not* be an appropriate criterion for deciding whether the atoms in adjacent lattice planes can be resolved. It would be appropriate only if the electrons scattered by these atoms had a random phase relation. In fact, the electrons scattered by adjacent planes of atoms have a consistent phase relation, and can therefore interfere with one another. Diffraction effects by the crystal lattice are thus of paramount importance. We therefore start with equations 6.38 and 6.40, which give the distribution of the electrons of the direct and diffracted beams at the exit surface of a parallel-sided crystal of thickness D according to the two-beam approximation of the dynamical theory. If we suppose the crystal lamina to be approximately perpendicular to the electron beam, we can write

$$\alpha \approx 0 \qquad \alpha' \approx 0$$
$$\sec \alpha = 1 \qquad \cos \alpha = \cos \alpha' = 1$$
$$\sin \alpha' - \sin \alpha \approx \alpha' - \alpha.$$

Reference to Fig. 6.2(a) shows that $\alpha' - \alpha$ is the angle between the incident and diffracted rays, and is therefore equal to 2θ, where θ is the Bragg angle. Using the Bragg relation, we can therefore write

$$\sin \alpha' - \sin \alpha \approx \lambda/d$$

so that

$$k_0(\sin \alpha' - \sin \alpha)\xi = 2\pi\xi/d. \qquad 10.9$$

With these approximations, we can remove a common factor

$$\exp \{i(k_0 + \pi g \varepsilon)D\} \exp \{ik_0\xi \sin \alpha\}$$

from 6.38 and 6.40, and obtain for the complex amplitudes of the two beams

$$G = \frac{\sin \pi SD}{\Delta S} \exp \{i\pi/2\} \qquad 10.10$$

and

$$T = \left[1 - \frac{\sin^2 \pi SD}{\Delta^2 S^2}\right]^{\frac{1}{2}} \exp \{i(2\pi\xi/d - \sigma)\} \qquad 10.11$$

where

$$\tan \sigma = \left[1 - \frac{1}{\Delta^2 S^2}\right]^{\frac{1}{2}} \tan \pi SD. \qquad 10.12$$

The intensity distribution of the electrons at the exit face is therefore

$$|G + T|^2 = 1 + \sin 2\tau \sin (2\pi\xi/d - \sigma), \qquad 10.13$$

where

$$\sin \tau = \frac{\sin \pi SD}{\Delta S}. \qquad 10.14$$

The expression 10.13 has a periodicity in ξ equal to d, the spacing of the atomic planes in the crystal. Provided the resolution of the microscope is adequate, we shall therefore observe the image of the crystal crossed by fringes having a sinusoidal intensity distribution. The separation of these fringes, referred to the object plane, is equal to the spacing of the atomic planes. In this limited sense, we may claim to 'see' the lattice planes in the microscope.

A necessary condition for the fringes to appear in the image is, of course, that both the direct and diffracted beams shall enter the objective aperture. The angle between these beams is twice the Bragg angle, and is typically about 2×10^{-2} radian. This requires the objective aperture to accept a cone of rays of semi-angle greater than the optimum angle of 6×10^{-3} radian quoted above. Spherical aberration will therefore be of crucial importance in deciding whether the lattice planes are resolved in the image; we shall discuss this point in more detail later.

It must be emphasized that even if the resolution of the microscope suffices to make visible the fringes described by the expression 10.13, this does not mean that the actual lattice planes are observed. The expression 10.13 shows that both the position and visibility of the fringes depend on the thickness of the crystal and the amount ε by which the beam incident on the crystal deviates from the exact

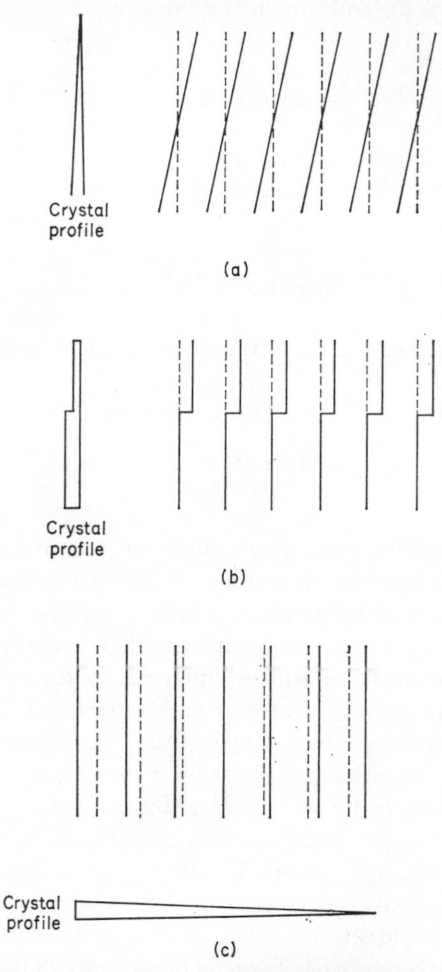

Fig. 10.6. Distortion of lattice plane 'images' formed by crystals of non-uniform thickness. In each case, the dotted lines indicate the fringes which would be formed by a crystal of uniform thickness; the full lines show the fringes formed by a crystal of non-uniform thickness.

(a) Crystal tapers in a direction parallel to the fringes, resulting in oblique fringes.
(b) Crystal with a step, resulting in a jog in the fringe pattern.
(c) Crystal tapers in a direction perpendicular to the fringes, resulting in a change in the fringe spacing.

Bragg angle. In particular, we note that if the crystal thickness D is an integral multiple of $1/S$, then $\sin 2\tau = 0$ and the fringes are invisible; the fringes have maximum visibility for half-integral values of SD. The position of the fringes is determined by the term σ in 10.13 and therefore, by virtue of 10.12, by the value of SD. We note that an increment of the crystal thickness D of $1/S$, i.e. (cf. 6.37) approximately the extinction distance Δ, will increase σ by π; this will change bright fringes to dark and vice versa.

The dependence of the position of the fringes on the thickness of the crystal results in a distortion of the fringe pattern if the crystal is not of uniform thickness. For example, if the crystal tapers in a direction parallel to the fringes (Fig. 10.6(a)) portions of the same fringe will be displaced laterally by different amounts (due to the differing values of σ), and the fringes will be inclined to the direction of the lattice planes. In the same way (Fig. 10.6(b)) a step-wise change of thickness will produce a jog in the fringes. On the other hand (Fig. 10.6(c)), if the crystal tapers in a direction normal to the fringes, successive fringes will have different values of σ and will be displaced by different amounts. Thus the fringe spacing will not be exactly equal to the lattice plane spacing. These effects have all been observed experimentally (Chadderton [12]). Obviously, the interpretation of lattice plane 'images' requires great caution.

If diffracted beams from *two* sets of planes inclined to one another can enter the microscope objective, then we may expect a pattern arising from the superposition of two inclined sets of fringes. This will appear as a regular pattern of spots which will represent the projection of the atoms of the crystal on to a plane normal to the incident electron beam.

It will have been observed in the foregoing discussion that the angular aperture β which the microscope objective must have in order to admit both the main and the diffracted beams is larger than the optimum value given in equation 10.7, which relates to the resolution of adjacent parts of an *aperiodic* structure. This leads us to reconsider the effect of spherical aberration on the image of a *periodic* structure. The essential difference between the two cases is that an aperiodic object will produce scattered rays which will fill the whole cone of semi-angle β: a periodic structure, on the other hand, will produce a few narrow, well-collimated beams (the direct and diffracted beams) filling only a small fraction of the object aperture.

If the main beam is directed along the axis of the objective, and the diffracted beam makes an angle β with this axis, then it can be shown that the effect of spherical aberration is to introduce a phase difference.

$$\phi = \frac{\pi C_s \beta^4}{2\lambda} \qquad 10.15$$

between the two beams. This, by itself, will not limit the resolution of the fringes. A little consideration of equations 10.10–10.13 will show that the only effect of this phase difference is to shift the lattice fringes, a phase difference of 2π corresponding to a displacement of one fringe spacing. However, if the diffracted beam is not perfectly collimated, but has an angular spread $\delta\beta$, then there will be a corresponding spread in the phase difference of

$$\delta\phi = 2\pi C_s \beta^3 \delta\beta / \lambda \qquad 10.16$$

and a broadening of the fringes by a fraction

$$\delta\phi / 2\pi = C_s \beta^3 \, \delta\beta / \lambda$$

of a fringe spacing. It is reasonable to suppose that the fringes will just cease to be resolved if this broadening amounts to half the spacing between the fringes. Then the maximum permissible value of $\delta\beta$ will be

$$\delta\beta = \lambda / 2C_s\beta^3. \qquad 10.17$$

The angular spread $\delta\beta$ of the diffracted beam is due to (*a*) the angular spread $\delta\varepsilon$ of the primary beam, and (*b*) the finite size of the specimen. The angular spread due to (*b*) may, with sufficient accuracy, be written as λ/w, where w is the width of the specimen (equation 3.22b). Since β, the angle between the direct and diffracted beam, is twice the Bragg angle, we may, with sufficient accuracy, write it as λ/d, where d is the spacing of the planes producing the fringes. On inserting these values of β and $\delta\beta$ in 10.17, we find for the smallest value of d, i.e. the smallest lattice plane spacing which can be resolved

$$d_{\text{min}} = [2C_s\lambda^2(\delta\varepsilon + \lambda/w)]^{\frac{1}{4}}. \qquad 10.18$$

The resolution limit therefore depends on the width of the specimen.

Using the typical values quoted in 10.8, we find from 10.18 that to resolve lattice plane fringes of 200 pm separation would require an angular spread $\delta\varepsilon$ of the primary beam of not more than about 10^{-5} radian and a specimen width of about 200 nm. The limit for $\delta\varepsilon$ can just be met with conventional electron microscopes at the expense of a reduction of image brightness. To meet the requirement on w needs a careful selection of the specimen. Reference to 10.12 and 10.13, which give the dependence of the position of the fringes on the specimen thickness D, shows that the specimen thickness must be constant to within a fraction of the extinction distance Δ over the required width of the specimen.

Appendices

A.1 *Relation between electronic scattering factor and Fourier potential coefficient*

The discussion in § 3.3 is based on the concept of the crystal as an array of atoms, each of which scatters a wave

$$\psi = -f\, e^{i\mathbf{k}\cdot\mathbf{r}}/r, \qquad 3.1$$

where
$$f = e(Z - F)/(16\pi\varepsilon_0 P^* \sin^2 \theta). \qquad 3.3$$

The first term in the expression for f represents the scattering by the nucleus, a point charge $+Ze$. The second term represents the scattering by Z extra-nuclear electrons. The factor F is a function of $(\sin \theta/\lambda)$, and takes account of the phase difference between waves scattered by different parts of the atom.

In contrast to the above, the calculation of the intensity of a diffracted wave on the basis of the dynamical theory in § 6.2 starts by regarding the crystal as a three-dimensional, periodic potential field. This, in turn, implies that the crystal is a three-dimensional, periodic distribution of charge. We now combine with this concept the idea implied in 3.3 that a point charge q scatters a wave given by 3.1 with

$$f = q/(16\pi\varepsilon_0 P^* \sin^2 \theta).$$

This leads to the result that the amplitude of the wave scattered by the whole crystal in a direction corresponding to the reciprocal lattice vector $\mathbf{g}$ can be written as

$$-\frac{e^{i\mathbf{k}\cdot\mathbf{r}}}{r}\,\frac{1}{16\pi\varepsilon_0 P^* \sin^2 \theta}\int_{\text{crystal}} e^{2\pi i\mathbf{R}\cdot\mathbf{g}}\,\rho\, d\tau,$$

where $\rho\, d\tau$ is the charge in a volume element $d\tau$ situated at a point whose position vector is $\mathbf{R}$, and $\mathbf{g}$ is given by 3.7. The exponential factor in the integrand takes account of the phase difference between waves scattered from different parts of the crystal.

This integral over the whole volume of the crystal can be written as the product of an integral over a unit cell and the lattice factor G

(cf. equations 3.9 and 3.10). Hence we may write for the scattered amplitude

$$-\frac{e^{i\mathbf{k}\cdot\mathbf{r}}}{r}\frac{G}{16\pi\varepsilon_0 P^* \sin^2\theta}\int_{\text{cell}} e^{2\pi i\mathbf{R}\cdot\mathbf{g}}\,\rho\,d\tau.$$

If we express the charge density ρ as a Fourier series

$$\rho = \sum \rho_{\mathbf{g}}\, e^{-2\pi i\mathbf{R}\cdot\mathbf{g}} \qquad \text{A.1.1}$$

then

$$\int_{\text{cell}} e^{2\pi i\mathbf{R}\cdot\mathbf{g}}\,\rho\,d\tau = \rho_{\mathbf{g}} v$$

where v is the volume of the unit cell, and the scattered amplitude is

$$-\frac{e^{i\mathbf{k}\cdot\mathbf{r}}}{r}\frac{G\rho_{\mathbf{g}} v}{16\pi\varepsilon_0 P^* \sin^2\theta}. \qquad \text{A.1.2}$$

Now the electrostatic potential V within the crystal is related to the charge density by Poisson's equation

$$\nabla^2 V = -\rho/\varepsilon_0 \qquad \text{A.1.3}$$

If we write the potential as a Fourier series

$$V = \sum V_{\mathbf{g}}\, e^{2\pi i\mathbf{R}\cdot\mathbf{g}}, \qquad \text{A.1.4}$$

then, substituting A.1.1 and A.1.4 in A.1.3

$$4\pi^2 g^2 V_{\mathbf{g}} = \rho_{\mathbf{g}}/\varepsilon_0. \qquad \text{A.1.5}$$

Equating the expressions A.1.2 and 3.10a, and introducing A.1.5, we find that

$$V_{\mathbf{g}} = \frac{E}{v}\frac{4P^* \sin^2\theta}{\pi g^2}.$$

Using the Bragg equation

$$2d\sin\theta = \frac{2\sin\theta}{g} = \lambda;$$

this becomes

$$V_{\mathbf{g}} = \frac{EP^*\lambda^2}{v\pi}. \qquad \text{A.1.6}$$

A.2 *Derivation of Bragg's equation*

A strong reflection occurs from a set of parallel planes of atoms if the path difference between rays reflected at successive planes is an

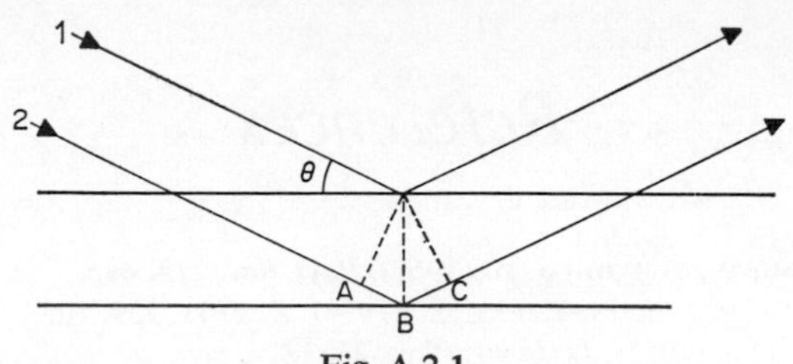

Fig. A.2.1.

integral number of wavelengths. In Fig. A.2.1, the path difference between the rays 1 and 2 is clearly

$$ABC = 2d' \sin \theta,$$

where d' is the interplanar spacing and θ is the glancing angle of the incident and reflected beams at the atomic planes. Hence the condition for a strong reflection is

$$2d' \sin \theta = n\lambda, \quad \text{A.2.1}$$

where n is an integer. It is usual to write this equation in the form

$$2d \sin \theta = \lambda, \quad \text{A.2.2}$$

where $$d = d'/n \quad \text{A.2.3}$$

is a fictitious interplanar spacing equal to the true spacing divided by the order of the interference.

References

1. AHARONOV, Y., and BOHM, D. (1959). *Phys Rev.* **115,** 485.
2. ALTENHEIN, H. J. and MOLIÈRE, K. (1954). *Z. Phys.* **139,** 103.
3. BARTELL, L. S. (1955). *J. Chem. Phys.* **23,** 1219.
4. BARTELL, L. S., BROCKWAY, L. O. and SCHWEDEMAN, R. H. (1955). *J. Chem. Phys.* **23,** 1854.
5. BASSETT, G. A. and KELLER, A. (1964). *Phil. Mag.* **9,** 817.
6. BEWILOGUA, L. (1931). *Phys. Z.* **32,** 740.
7. BROCKWAY, L. O. and BARTELL, L. S. (1954). *Rev. scient. Instrum.* **25,** 569.
8. BROCKWAY, L. O. and WALL, F. T. (1934). *J. Am. Chem. Soc.* **56,** 2373.
9. BUHL, R. (1959). *Z. Phys.* **155,** 395.
10. DE BROGLIE, L. (1924). *Phil. Mag.* **47,** 446.
11. CALDWELL, C. W. (1965). *Rev. scient. Instrum.* **36,** 1500.
12. CHADDERTON, L. T. (1961). *Nature, Lond.* **189,** 564.
13. COSSLETT, V. E. (1946). *Introduction to Electron Optics*, Clarendon Press, Oxford.
14. COWLEY, J. M., GOODMAN, P. and REES, A. L. G. (1957). *Acta Crystallogr.* **10,** 19.
15. COWLEY, J. M. and REES, A. L. G. (1947). *Proc. phys. Soc.* **59,** 287.
16. DAVISSON, C. and GERMER, L. H. (1927). *Phys. Rev.* **30,** 705.
17. DEBYE, P. (1939). *Phys. Z.* **40,** 573.
18. DEBYE, P. (1941). *J. Chem. Phys.* **9,** 55.
19. DITCHBURN, R. W. (1963). *Light*, Blackie, London.
20. DRUMMOND, G. D. (1950). *J. R. Microsc. Soc.* **70,** 1.
21. FARNSWORTH, H. E. (1929). *Phys. Rev.* **34,** 679.
22. FARNSWORTH, H. E. (1950). *Rev. scient. Instrum.* **21,** 102.
23. FARNSWORTH, H. E., SCHLIER, R. E., GEORGE, T. H. and BURGER, R. M. (1955). *J. appl. Phys.* **26,** 252.
24. FINCH, G. I., QUARRELL, A. G. and WILMAN, H. (1936). *Trans. Faraday Soc.* **31,** 1051.
25. GERMER, L. H. and MACRAE, A. U. (1962). *J. appl. Phys.* **33,** 2923.
26. GERMER, L. H., MACRAE, A. U. and HARTMAN, C. D. (1961). *J. appl. Phys.* **32,** 2432.
27. GRIGSON, C. W. B., DOVE, D. B. and STILWELL, G. R. (1965). *Nature, Lond.* **205,** 1198.
28. GRIGSON, C. W. B. and TILLETT, P. I. (1968). *Int. J. Electron.* **24,** 101.
29. HAINE, M. E. and EINSTEIN, P. A. (1952). *Br. J. appl. Phys.* **3,** 40.
30. HASHIMOTO, H., HOWIE, A. and WHELAN, M. J. (1962). *Proc. R. Soc. A.* **269,** 80.
31. HIRSCH, P. B., HOWIE, A., NICHOLSON, R. B., PASHLEY, D. W. and WHELAN, M. J. (1965). *Electron Microscopy of Thin Crystals*, Butterworths, London.
32. HONJO, G., KODERA, I. and KITAMURA, N. (1964). *J. Phys. Soc., Japan* **19,** 351.

33. IBERS, J. A. and VAINSHTEIN, B. K. (1962). *International Crystallographic Tables* 3, Kynoch Press, Birmingham.
34. KARLE, J. and HAUPTMAN, H. (1950). *J. Chem. Phys.* **18,** 875.
35. KAY, D. H. (1965). *Techniques for Electron Microscopy*, Blackwells, Oxford.
36. KELLER, M. (1961). *Z. Phys.* **164,** 274.
37. KIENDL, H. (1967). *Z. Naturforsch.* **22a,** 79.
38. KITAMURA, N. (1966). *J. appl. Phys.* **37,** 2187.
39. KOMATSU, K. (1964). *J. Phys. Soc., Japan* **19,** 1243.
40. LANDER, J. J. and MORRISON, J. (1963). *J. appl. Phys.* **34,** 1403.
41. LEBEDEFF, A. A. (1931). *Nature, Lond.* **128,** 491.
42. LE POOLE, J. B. (1947). *Philips techn. Rundsch.* **9,** 33.
43. MACRAE, A. U. (1964). *Surface Sci.* **1,** 319.
44. MARTON, L., SIMPSON, J. A. and SUDDETH, J. A. (1954). *Rev. scient. Instrum.* **25,** 1099.
45. MENTER, J. W. (1950). *J. scient. Instrum.* **27,** 335.
46. MEYERHOFF, K. (1959). *Acta Crystallogr.* **12,** 330.
47. MOLIÈRE, K. and ALTENHEIM, H. J. (1954). *Z. Phys.* **139,** 103.
48. MOLIÈRE, K. and NIEHRS, H. (1954). *Z. Phys.* **137,** 445.
49. MOLIÈRE, K. and NIEHRS, H. (1955). *Z. Phys.* **140,** 581.
50. MÖLLENSTEDT, G. and BAYH, W. (1962). *Naturwiss.* **49,** 81.
51. MÖLLENSTEDT, G. and DÜKER, H. (1956). *Z. Phys.* **145,** 377.
52. MORSE, P. M. (1932). *Phys. Z.* **33,** 443.
53. MOTT, N. F. and MASSEY, H. S. W. (1965) *The Theory of Atomic Collisions*, Clarendon Press, Oxford.
54. NISHIKAWA, S. and KIKUCHI, S. (1928). *Nature, Lond.* **121,** 1019.
55. PARK, R. L. and FARNSWORTH, H. E. (1964*a*). *Rev. scient. Instrum.* **35,** 1592.
56. PARK, R. L. and FARNSWORTH, H. E. (1964*b*). *Surface Sci.* **2,** 527.
57. PHILLIPS, F. C. (1946). *An Introduction to Crystallography*, Longmans, Green, London.
58. PICARD, R. G., SMITH, P. C. and REISNER, J. H. (1949). *Rev. scient. Instrum.* **20,** 601.
59. RIECKE, W. D. (1961). *Optik* **18,** 278.
60. RYMER, T. B. and WRIGHT, K. H. R. (1952). *Proc. R. Soc. A* **215,** 550.
61. SCHEIBNER, E. J., GERMER, L. H. and HARTMAN, C. D. (1960). *Rev. scient. Instrum.* **31,** 112.
62. STOKES, A. R. and WILSON, A. J. C. (1942). *Proc. Camb. Phil. Soc.* **38,** 313.
63. THOMSON, G. P. and REID, A. (1927). *Nature, Lond.* **119,** 890.
64. TRÜB TÄUBER (1955). *J. scient. Instrum.* **32,** 407.
65. VAINSHTEIN, B. K. (1964). *Structure Analysis by Electron Diffraction*, Pergamon Press, Oxford.
66. WIERL, R. (1931). *Ann. Phys., Germany* **8,** 521.
67. WITT, W. (1964). *Z. Naturforsch.* **19a,** 1363.
68. ZWORYKIN, V. K., MORTON, G. A., RAMBERG, E. G., HILLIER, J., and VANCE, A. W. (1945). *Electron Optics and the Electron Microscope*, John Wiley, New York.

Index